Progressive Beef Cattle Raising

by Edward Norris Wentworth

with an introduction by Jackson Chambers

This work contains material that was originally published in 1920.

This publication is within the Public Domain.

This edition is reprinted for educational purposes and in accordance with all applicable Federal Laws.

Introduction Copyright 2018 by Jackson Chambers

Self Reliance Books

Get more historic titles on animal and stock breeding, gardening and old fashioned skills by visiting us at:

http://selfreliancebooks.blogspot.com/

Introduction

I am pleased to present another title in the "Cattle" series.

The work is in the Public Domain and is re-printed here in accordance with Federal Laws.

As with all reprinted books of this age that are intended to perfectly reproduce the original edition, considerable pains and effort had to be undertaken to correct fading and sometimes outright damage to existing proofs of this title. At times, this task is quite monumental, requiring an almost total "rebuilding" of some pages from digital proofs of multiple copies. Despite this, imperfections still sometimes exist in the final proof and may detract from the visual appearance of the text.

I hope you enjoy reading this book as much as I enjoyed making it available to readers again.

Jackson Chambers

Table of Contents

	PAGE
INTRODUCTION, by James Brown	6
THE FUNCTION OF CATTLE	9
Position of Cattle in the System of Farming	9
Origin and Kinds of Cattle	9
CATTLE BREEDS	11
Breed Qualifications	11
The Purebred Animal	11
How the Purebred Developed	12
The Pedigree	13
CATTLE BREEDING	15
How Cattle Are Improved	15
Grading Up Beef Cattle	15
The Relative Influence of Sire and Dam	17
The Proportion of Purebred Cattle	17
Community Breeding	18
The Distribution of the Breeds	19
THE PRODUCTION OF BEEF CATTLE	20
Foundation Blood for Beef Production	20
The Problems of the Range Cattle Breeder	20
Buying Feeders	21
Feeding Equipment	21
Some Cattle Rations	23
Growing the Calf	25
The Advantage of Young Cattle	27
Essentials of a Complete Ration	28

	PAGE
Silage	29
Requisites of a Good Silo	30
Silo Capacities	30

MANAGEMENT OF THE BEEF HERD ... 32

Three Types of Cattle Farming	32
The Maintenance of the Breeding Herd	33
The Pasture	34
The Contents of the Hay Stack	34
Sanitation on the Farm	36
Cattle Diseases	37
The Cow and Her Calf	41
Gestation Table	41

THE CATTLE INDUSTRY ... 43

The United States' Position in Beef Production	43
The American Beef Export Trade	44
Relation of Export Trade to Cattle Production	44

CATTLE PRICES ... 46

The Relation of the Market to the Feeding Business	46
Why Markets Fluctuate	47
The Two Classes of Price Fluctuations	47
Seasonal Variations in Price	48
The Problem of Marketing Beef in All Seasons	49
Methods of Reaching the Most Favorable Markets	49
The Effect of Supply and Demand on Hoof Prices	50

THE BEEF CARCASS ... 52

The Relative Value of Carcass Cuts	52
Factors in Carcass Values	53
The Relation of Carcass Price to Hoof Price	54

	PAGE
MARKET CLASSES OF CATTLE	56
How Cattle Are Classified	56
How Cattle Are Graded	57
Characteristics of Different Grades and Classes of Beef Cattle and Butcher Stock	58
Grades and Classes of Feeders and Stockers	59
CATTLE TYPES	61
How Type Is Determined	61
Characteristics of the Standard Types of Beef Steer	62
Dressing Percent	62
MARKETING CATTLE	64
Preparations for Shipping	64
Shipping Counsel	64
Handling Cattle at the Market	66
Slaughtering Cattle	67
Byproducts	69
REFERENCES	71

Introduction

THE growing tendency among cattle feeders is to regard quality and finish in cattle as superfluous, due to the narrowing margin in price between the poorest kinds of cattle that reach the market and those of best grade. The majority of feeders can remember when such animals as canner cows had no value with the packer or retail butcher, and the increasing uses which have been found for them, sufficient to give quotable prices from day to day, have been interpreted by these feeders to mean that quality no longer has the value it once enjoyed. Nothing could be further from the truth. Quality cattle will always be properly appraised, because they produce the class of meat that is easiest to sell, requiring a minimum of effort on the part of the ultimate salesman.

From year to year the standards as to market types and classes are changing, based on the changing demands of the consumer. The cattle feeder usually learns of these changes when he sells, and occasionally feels that the market asks for any kind of cattle other than what he brings. The chief difficulty in meeting exactly the market demands, lies in the fact that the standards of cattle of one or two decades ago still persist in the minds of many feeders and are perpetuated by the types of steers recognized by the majority of judges in the fat stock shows. Whether or not the feeder intends to do so, he carries in his mind the standard of perfection established by the heavy, richly finished bullocks popular twenty years ago, and he interprets the trimmer killing characteristics which modern cattle show, coupled with lesser size, as distinct steps backward.

The chief factor in bringing about the change in type has been the change in retail demand. The public, while more fastidious as to the cuts of beef it consumes, does not eat as much meat as it did formerly, and will not tolerate the waste in cuts that the rich steaks and roasts of a half century ago possessed. The modifications in market standards are based on these two simple facts, and the trade must educate itself to the idea. The retailer has been most sensitive to this change in demand, but the reaction on the packer has been so direct that he has been forced to translate immediately the desires of the consumer into a type of cattle suitable for the production of the best selling cuts. The principal factors that have caused the change are the increase in the population of cities coupled with the reduced ratio of producers; the inroads on the family purse made by luxuries, which have restricted the percentage spent on necessities; and the reduced size of families which has permitted groceries with small meat shops vending pound to two-pound cuts to make deep inroads into the business of the specialized butcher.

The demand for heavy cattle varies little throughout the year. The markets of the big cities, principally New York, Chicago, Philadelphia and Boston, supply the principal trade, the sales being largely to hotels and clubs that have a standard demand for certain cuts the year around. This natural demand for heavy finished cattle takes only about 15 per cent of the cattle on the market, their live weight being 1300 pounds and up, and their carcasses making about 750 pounds of beef and up. In order that prices for this class of cattle remain steady and hold the same relative relation to other classes of cattle, the supply must be regular as a few too many can readily glut the market. Unfortunately, it is difficult to finish cattle of this sort at all seasons of the year; few of them coming on the market in the period from August 1st to February 1st excepting cattle finished for shows and for the Christmas trade. From February on there is usually a sufficiency of this class of cattle, while late March and April may find a few too many with consequent drops in price. On the other hand, in times of scarcity of heavy cattle, buyers having orders for this class of stock make competition so lively that steers weighing 1400 pounds or up which will satisfy their trade, often bring $1 to $2 above their real value as compared to smaller cattle of the same grade. A judgment of values based on the price of this class of cattle either in times of scarcity or surplus, is bound to be misleading to the average feeder.

The profit in beef production in the future will increasingly lie in quality stock. Early maturity and quick money turn-overs are certain to be the keynotes of future meat production, due to high land and feed values. Cold blooded stock will never utilize feed for fattening and finishing until the animals are well grown, three years old or over, while breeders and feeders will need to have their money back out of their animals by the time they are two years old, unless the cattle are range grown, when the difference in production costs will permit their profitable retention for another year. A six-months calf of good blood, dropped in the spring, can be fattened so as to market it the following spring or summer from any cornbelt farm, but a six-months scrub will not efficiently utilize its feed because its growth is slow, and it will not develop as rapidly, fatten as well, nor grow as satisfactorily as the well-bred animal. Well-bred calves can always be finished from calfhood on, and can make the best quality of carcass, since they can utilize efficiently feed that the scrub cannot consume economically from lack of capacity or from inability to fatten or grow.

The use of purebred sires is the certain means of success in the future. This does not mean the indiscriminate use of such animals without regard to results—prices paid for them must be always as firmly grounded in returns as prices paid for feeders—but it does indicate that through them the efficient beef production of the future must be built. Quality as recognized today means better meat for the consumer, better killing qualities for the packer, and more efficient feeders for the producer. People are not buying meat nowadays to throw part of it away, and the majority of families with restricted pocketbooks are buying the medium weight cuts. The livestock market simply interprets the tendency of the meat-eating public and it recognizes that consumers will not tolerate waste except at a discount.—JAMES BROWN.

PROGRESSIVE BEEF CATTLE RAISING

Part I.

The Function of Cattle

Position of Cattle in the System of Farming

Beef cattle are the keystone of live stock farming. They fit into the farm economy more perfectly than any other animal because they require less labor for their care, they are less subject to disease, they consume cheap roughages and high priced concentrates in proportions better suited to ordinary farm rotations, and their product is less subject to speculative and seasonal fluctuations than any other class of meat animals. Beef cattle take less fertility from the farm when they are marketed than any other major farm product. Grain can be sold for cash as can the better quality of hay, but low grade hay, cane, corn stalks, and pasture have a very limited and unremunerative market. The direct sale of any of them increases the costs in harvesting and marketing and reduces the producing value of the land to such an extent that few farmers can get a proper return for it. Cattle supply the necessary market for both crops and farm labor, and through manure retain or even increase the richness, mellowness, life and waterholding capacity of the soil. Cattle feeding is the ideal operation from the standpoint of permanence of farming, of acquiring a comprehensive farm equipment, and of fully employing farm labor the year around.

Origin and Kinds of Cattle

Cattle were the first animals domesticated by man for purely agricultural purposes. They were kept for their meat and hides only in the earlier times, but later were milked and still later used for draft purposes. Unfortunately the kind of cattle that are best for beef are not best for

PROGRESSIVE BEEF CATTLE RAISING

milk or draft, and for many centuries cattlemen have selected their animals with these differences in mind. The breeders of continental Europe have tried to combine in their breeds all of the traits that make animals useful for milk, beef and draft, so-called triple purpose animals, but since many of the characters are antagonistic to each other, certain compromises in type have had to be made, which have rendered the animals less efficient for each of the special purposes. The principal example of this type is the Fribourgeois, the popular yellow and white breed of northeastern France, southeastern Belgium, Luxemburg and southwestern Germany. The almost complete replacement for draft purposes of cattle by horses in Great Britain and America many years ago has rendered the draft requirement unnecessary, and the common breeds in use in these two countries are either dual purpose or single purpose. The dual purpose breeds are those that have a fair value both for milk and beef, and include the Milking Shorthorn, the Red Poll, and the Devon. The special purpose breeds are either milk or beef, although each dairy breed has some value as a meat animal and each beef breed has some value as a milk producer. The dairy breeds in order of their beef value are Brown Swiss, Holstein-Friesian, Ayrshire, French Canadian, Guernsey and Jersey. The beef breeds are led decidedly by the Shorthorns and Polled Shorthorns in milk production, with the Herefords, Polled Herefords, Aberdeen-Angus and Galloways, secondary in this particular.

Part II.

Cattle Breeds

Breed Qualifications
As shown in the preceding paragraph, beef cattle may come from any of six specialized breeds, while dual purpose and even dairy breeds show some beef merit. Within the beef breeds themselves, there is more difference between good and poor animals of one breed than there is between the different breeds. Supporters of each breed claim special characteristics for their favorites which are supposed to make them better than their rivals, but it has never been proved that the qualities for which one breed may be famous do not appear in the representatives of other breeds. For example, rustling qualities and ability to fatten on grass are supposed to be pre-eminent Hereford characteristics; quality, milk production and adaptability, Shorthorn characteristics; and wonderful hardiness, a Galloway characteristic; yet records of our fairs and ranges can nearly always show where the breed in question has been excelled by a few individuals of other breeds in the points where supremacy was claimed. Hence the young breeder can well afford to take the stock that suits his purpose where he is located, regardless of breed, keeping always in mind the market for his young animals.

The Purebred Animal
Race horse breeders have a maxim that "What is bred in the bone will come out in the flesh." This applies in a striking way to cattle, for "cat-hammed, fish-backed" dams and sires produce "cat-hammed, fish-backed" calves just as surely as one rolling tumbleweed will infect a field. Conversely, broad-backed, meaty cattle will as certainly produce broad-

PROGRESSIVE BEEF CATTLE RAISING

backed meaty calves. This quality of being able to transmit to the offspring the beef qualities for which selection has been made, convinced breeders that selected stock was purer in its inheritance than unselected and the term *purebred* was adopted for such animals. This does not mean that the animals are absolutely pure for such traits and will transmit no others, but it does mean that they will do so with infinitely greater regularity than the unselected kind. No strain of cattle now exists that cannot be made to breed more uniformly than it does now, but the average purebred animal breeds so much more uniformly than his grade and scrub rivals that his worth cannot be gainsaid. The cattle raiser who uses purebred bulls over a period of years invariably has more uniform and cheaper finishing steers than the man who uses bulls of mixed bloods.

How the Purebred Developed

Pure breeds of beef cattle arose through the selection of animals slightly superior to the stock of the surrounding districts with respect to beef production. The first improver of beef cattle was an Englishman, a resident of Leicestershire, named Robert Bakewell. He worked with the old Longhorn stock of central England, and being a skilled anatomist was able to appreciate the means whereby changes in external form would affect the carcass. He selected for increased thickness of loin, rib and quarter, for more rapid fattening qualities, and for early maturity. By mating together related animals he fixed these traits so strongly that his cattle became known all over England, while his sheep which he improved by similar methods, were so well known that George Washington imported rams of Bakewell breeding for use on his Mt. Vernon estates. From a careful study of Bakewell's methods the Colling brothers, the Booth and the Bates families established Shorthorn cattle, and a few years later the Tompkins, Prices and Hewers founded the Herefords.

PROGRESSIVE BEEF CATTLE RAISING

Three or four decades after this the foundations of the Aberdeen-Angus were securely laid by Hugh Watson in the north of Scotland, while numerous breeders were busily evolving the Galloway. During the eighties of the last century the problem of horn bruises in shipping cattle became so important in America that the Polled Shorthorn and Polled Hereford arose. Each of these breeds was developed to meet a special economic need, and the animals chosen as founders of the breed were arbitrarily selected on the basis of their transmitting the desired characters to their descendants.

The Pedigree

The ultimate test of the purebred animal is its possession of a registered pedigree. Not the mere possession of a pedigree counts, for all cattle have pedigrees. This term is just another name for ancestors, or a tabulation of ancestors. All cattle have ancestors, but purebred cattle have certified ancestors, and it is well known that the average of its ancestors are well above the average of the ancestors of scrub animals. Poor cattle with poor cattle in their pedigrees produce poor cattle, while good cattle with good cattle in their pedigrees produce good cattle. Registered animals have a recorded pedigree behind them which shows the kind of stuff of which they are made. The registry association by its approval of the animal's ancestry gives assurance to the breeder of a well marked standard of calves resulting from its use.

PROGRESSIVE BEEF CATTLE RAISING

Idolmere 199904
- Oakville Quiet Lad 109210
  - Black Woodlawn 42088
    - Bell's Eclipser 26695
      - Moon Eclipser 8635
      - Belle of Cottage Grove 13150
    - Blackbird 13th 24464
      - Blackbird Jim 17564
      - Blackbird 8th 13190
  - Queen McHenry 47th 61884
    - Heather Blackbird 20333
      - Heather Lad 4th 16747
      - Blackbird of Turlington 6th 14484
    - Queen McHenry 5th 17490
      - Emu 1454
      - Queen Mother of Tillyfour 10093
- Home View Lady 86247
  - Pabmo 38977
    - Baltimore of Glendale 24275
      - Golden Abbott 16756
      - Pride of the Glen 12346
    - Pride McHenry 6th 23936
      - Heather Lad 4th 16747
      - Pride of Turlington 3d 12177
  - Lady Ideal 7th 20498
    - Black Aristocrat 11582
      - Abactur 7426
      - Blackbird of Hillhurst 2d 8048
    - Lady Ideal 3d 12330
      - Eros 5656
      - Lady Ideal 2d 7224

The tabulated pedigree of the $50,000 grand champion Aberdeen-Angus bull "Idolmere" at the 1919 International. This shows all ancestors of the bull back to the fourth generation, and furnishes the informed breeder knowledge of the probable breeding qualities of this bull.

PART III.
Cattle Breeding

How Cattle Are Improved

Native cattle in any section of the country have differed considerably from the type which has come to be accepted as standard by American steer feeders. The general procedure by which the quality of this native stock has been improved has been through the grading up with sires of improved and purebred stock. These animals come from strains which have been selected for special beef purposes and in addition to possessing the desired type are usually able to transmit it to their offspring. Despite the loyalty of the supporters of the different breeds to their favored stocks there is little difference in the ability of Shorthorns, Herefords, Aberdeen-Angus, and Galloways to transmit improved beef-making ability. Because the Aberdeen-Angus and Galloways happen to be black and polled their supporters have believed them to be more prepotent for grading purposes than some of the lighter colored animals but it is now known that these characteristics are inherited separately and have no relation necessarily to the transmission of fattening qualities and early maturity which are passed on on their own merits.

Grading up Beef Cattle

Grading up of the stock of a locality means literally to use purebred bulls on native cows generation after generation. The characters which make a purebred valuable are thus transferred to the herd, the degree of transfer depending on the number of crosses of improved blood. For example: If a Hereford be crossed to Florida "piney-woods" cows, half of the inherited characters of the offspring are Hereford, the other half, "piney-woods." If to these crossbred heifers, Herefords are again mated, on the average three quarters of the traits inherited will be Hereford, and one quarter "piney-woods." This does not mean that any single individual is three-fourths Hereford, but only that the average of all characters in that generation is three-fourths Hereford. As a matter of fact it is possible, although not probable, that some animals of the second cross might have entirely Hereford characters, with a similar number only half Hereford, and with the others intergrading between, but the average would still be three-quarters Hereford or 75 percent. By continuing the use of Hereford bulls for several generations, the proportion of Hereford characters would be increased, while some of the characters would be pure Hereford, beginning with the second cross. The proportion of these pure characters would increase with each generation although not as rapidly as the total number of Hereford characters. The more that Hereford sires are used, the more likely the resulting grade dams will transmit Hereford traits, and the purer the characters of the new generations of calves will be. The rate of increase in these characters is shown in the following table:

One cross.................50% Hereford traits.
Two crosses..............75% Hereford traits.
Three crosses............87.5% Hereford traits.
Four crosses.............93.75% Hereford traits.
Five crosses.............96.875% Hereford traits.
Six crosses..............98.4375% Hereford traits

And so on, each additional cross producing animals having a proportion of Hereford characters half way between the last generation and 100 percent. This does not apply to Herefords only but to any improved breed of livestock. Herefords are used simply for illustrative purposes. Some breeds seem to be more potent in transmitting their characters than others, but this is due to their possession of more noticeable characters than the others. For example the Aberdeen-Angus is black, a dominant color in inheritance no matter what breed of cattle it is found in, and polled, also a dominant character, hence it markedly affects the appearance of its calves. This is only in appearance, however, as other breeds are able to build up the fattening tendency just as rapidly.

The Relative Influence of Sire and Dam
Neither sire nor dam transmit to their offspring all of the qualities which they possess, but on the average each transmits only half. On this account it is highly important to have well bred bulls so that their contribution to the offspring will be more uniform. The cows of the herd always carry more widely divergent characters than it is possible for any one bull to possess, hence their hereditary contribution to the calves will be very much more variable than the hereditary contribution of the bull. Therefore the old statement that the bull is half of the herd is correct, but if we consider his power to make the herd uniform, we can really consider him as having more than a fifty percent influence.

The Proportion of Purebred Cattle
Purebred sires to increase the production of better beef animals, provide the most serious need of the livestock industry at present. While the cattle producers of the cornbelt and of the range, in general realize the desirability of the better bred sires there are other sections of the country in which the real value of the improved type is not understood, and in which such campaigns as the Better Sire Movement are bound to bring first-class results. Figures obtained from the secretaries of the different breed associations present the following facts for the consideration of beef cattle producers. Part of these figures are accurate while part are based on estimates, but they are the most reliable figures presented to date.

PROGRESSIVE BEEF CATTLE RAISING

Breed	Number Registered	Number Living	Number Unregistered	Number Living Bulls
Aberdeen-Angus	309,300	125,000	35,000	55,000
Herefords	889,000	500,000	....	....
Shorthorns	1,500,000	500,000	....	150,000
Galloways	47,300	25,000	....	....
Red Polls	95,900	18,500	6,000	4,000
Polled Herefords	21,000	12,000	7,000	7,000
Polled Shorthorns	48,400	25,000	6,500	11,000

On January 1, 1920, there were estimated by the United States Department of Agriculture to be 44,385,000 beef cattle in America. There are approximately 1,200,000 living registered purebreds reported in the preceding table or 2.7 percent of the total number of beef cattle in the country. The two societies having the largest number of unregistered purebreds, the Shorthorn and Hereford, were unable to estimate their respective numbers, but even if there were half as many unregistered as there are living registered animals the total number of purebreds in the United States would still be under 4 percent, a decidedly inadequate number from the standpoint of efficient beef animals.

Community Breeding One means whereby this low percentage of purebred animals can be more efficiently utilized is through the system of community or circuit breeding. If the farmers of a community organize to use one breed and exchange the sires either under a system of community ownership or by private sale to one another, additional years of service may be gotten out of high-class breeding bulls whose usefulness due to relationship to the females of the herd may be outlived in two or three years. Furthermore, the concentration of good cattle in a community will attract large numbers of purebred buyers and will give the market cattle shipped from the district a reputation among shippers and killers. Waukesha County, Wis., and Gage County,

PROGRESSIVE BEEF CATTLE RAISING

Neb., are well known examples of the advantage of co-operation between neighboring breeders.

The Distribution of the Breeds

While supporters of the different breeds of cattle try to claim all the good things in the beef category for their favorite breeds, the bulk of livestock men have a feeling that it is not a question of breed merit for beef production but of the merit of the individual bull, cow, or steer. Nevertheless the average run of the different breeds finds them definitely adapted to certain types of farming and certain general conditions, although no breed can be considered exclusively to monopolize a certain region or a certain function. For example: the Hereford is particularly well adapted to the range country because of its ability to fatten on grass; the Shorthorn is well adapted to general farming because of its size, ready fattening and milking qualities; the Aberdeen-Angus is especially well adapted to cornbelt feed lots and baby beef production because of its early maturity and most excellent carcass qualities; the Red Poll, to the central west because of its dual purpose ability; the Galloway, to the north and the short grass country because of its hardiness; and the Polled Shorthorns and Polled Herefords to the same regions as the horned breeds where hornlessness commands a premium. While individuals of each breed may succeed where other breeds are especially adapted, yet there is little doubt as to the general utility of the different breeds under the conditions described.

Part IV.

The Production of Beef Cattle

Foundation Blood for Beef Production

The success of growing cattle for the market depends in a large degree on the kind of calves that are produced Unless the right foundations in blood and type are laid, no amount of feeding by the professional feeder or skill in killing and cutting by the packer can make up for the original deficiency. Hence it is up to the breeder of feeding cattle to use the right kind of bulls and continually to breed up the females of his herd. Unless proper mating is made at the start, choice to prime steers are rarely if ever, produced.

The Problems of the Range Cattle Breeder

The producer of range cattle has many problems to face due to shortages of feed and water, that make his questions of special importance. The chief difficulty to engage his attention is the prevention of deterioration in size, and he must constantly introduce heavy boned bulls to keep his stock from "running out." It is the belief of students of this situation that the chief remedy to be applied is not through breeding but through feed, since the range is an abnormal environment for cattle bred to excel under the conditions of Scotland, England and the American cornbelt. The deficiency seems to be in the concentrates, that is, the protein and mineral matter, and for this purpose legumes should be introduced wherever possible, lespedeza, alfalfa, or any other clovers that make a good start. To tide over the feed shortages of snowbound winters and extra dry summers, it will pay the ranchman to grow corn, kafir, cane, or other crops that will make silage, and put it away in pit silos that can be dug at convenient spots for watering and feeding. These will be found very useful in ordinary seasons, and by keeping close watch for spoilage and using only the older silage, reserves may be maintained for three or four years. These may seem to involve a greater degree of extra work than the rancher would desire, but he must remember that only by the attention to such details as these, can he continue to make his business uniformly successful.

Buying Feeders

The feeder buyer has the most important job from the standpoint of profit and loss that is found in the cattle industry until the animals are finally bought for the kill. Few feeders realize the extreme importance of buying their animals at the right price. Fewer still realize the factors that determine feeder values. Until a few years ago feeder prices were determined by the cost of finishing, and cattlemen could take more chances on coming out through the big end of the horn than they can nowadays. Since the outbreak of the war feeds have increased in cost and the demand for meat has been so accelerated that the cheaper and lighter carcasses not salable in pre-war times are now taken at correspondingly high prices. In addition to the reaction against waste in cuts, there is an actual increase in the willingness to eat unfinished meats, so the spread between the thin and finished animal which held in past years has been decidedly narrowed. This necessitates the closest possible study of finishing costs on the part of the feeder buyer, and the most careful supervision of the purchase of his stock. Feeder cattle not prudently bought can never make money, no matter how economically the feeding may be managed.

Feeding Equipment

The majority of feeders do not recognize the importance of well arranged dry sanitary lots for steer feeding, nor do they realize the intimacy of the relation of conditions in the feedlot with the returns the steers are able to show. It is important from the standpoint of greatest efficiency in cattle feeding that not too many animals be confined in the same lot. The balance point as to numbers is determined by the rate at which the cattle gain and the increase in costs due to equipment and greater amount of labor. East of the Mississippi, the best results are obtained with from 30 to 45 steers in a lot, while farther west as many as 250 may be handled. The determining factors are the cost of labor and land as compared to the gains of the animals. The usual space required for steers has been found by experience to run about 90 to 100 square feet per head, including shed covering which should allow 20 to 25 square feet each. Where feeding is to be conducted over a period of time hard surface lots should be maintained, the shed being open, but closed to the north, west and east. Concrete foundations prove most suitable, but are too expensive in many sections of the country. Well packed crushed rock, brick or tile dust, or other surfacing material may be used, the latter being disadvantageous for cleaning although permitting the cattle to keep out of the muck. Drainage for cattle lots is perhaps most essential, as when the water runs off, even dirt lots may be kept in fair condition. Feed racks should have sufficient frontage to care for all of the steers without crowding. This means that there should be 2½ to 3 feet front per steer. Some feeders have had considerable success in using a combination feed rack for grain and roughage as indicated in Fig. A in the accompanying illustration, while others have a separate trough for grain and silage with a rack for hay as shown in Fig. B. The combination rack is a little cheaper to construct, but the separate racks operate a little more satisfactorily from the free choice basis for the steer, since the animals have a better chance to balance individually their grain against their roughage. Many steers come gradually to have their respective places at the trough, and although they do not consume all the grain, they keep other steers away by remaining to eat silage or hay. Feeding troughs should always be cleaned after each feeding and kept scrupulously sweet. Salt should always be available, as it promotes the animal's thrift and increases his ability to consume feed.

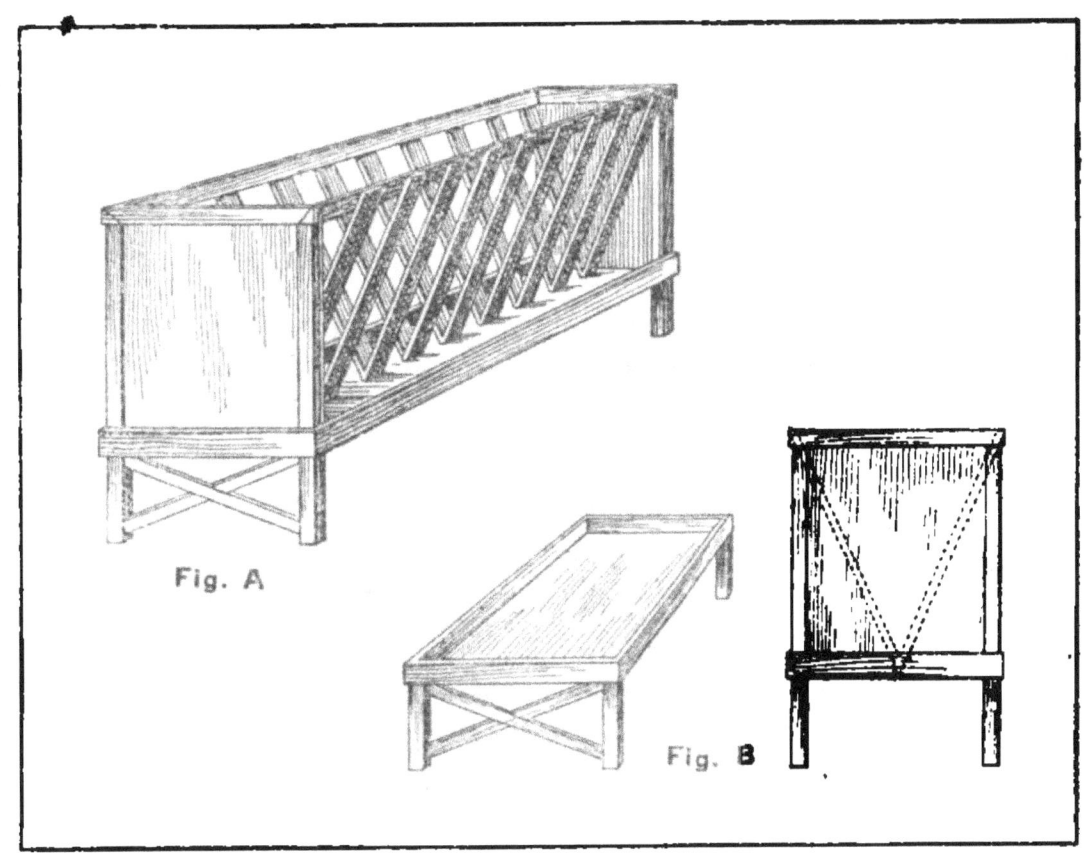

Some Cattle Rations

The great change in feed prices during the world war has to a certain extent invalidated the standard rations proposed by the experiment stations, and each state is now conducting investigations on the growing and feeding of cattle with cheap feeds, feeds that in smaller measure compete with human needs. This has limited very markedly the use of grains, so the following rations are composed of cheap and available feeds in the sections of the United States suggested:

PROGRESSIVE BEEF CATTLE RAISING

Section	Growing	Fattening
Bluegrass	Corn silage or Corn fodder Stalk Fields Straw stacks .5 lbs. oilmeal or cottonseed meal Bluegrass pasture	Corn silage 25-30 lbs. Clover or alfalfa hay 8-10 lbs. All corn steers will eat Oilmeal or cottonseed Oilmeal or cottonseed meal 1-2 lbs.
Southeast	Velvet bean fields Peanut hay in winter Cottonseed meal and hulls Bermuda grass pasture	Corn silage 25-30 lbs. Cottonseed meal 6-8 pounds or Velvet bean meal and cornstalk meal mixed all steers will eat
Central west	Corn silage or Corn fodder or Stalk fields Straw stacks Corn 3 to 6 lbs. in winter Oilmeal or cottonseed meal 1 to 2 lbs. Bluegrass pasture	Corn silage 25-30 lbs. Alfalfa dry clover hay 8-10 lbs. All corn steers will eat Oilmeal or cottonseed meal 1.5-2.5 lbs.
Northwest	Range Prairie or legume hay Straw stacks Sunflower silage Beet pulp where available	Damaged grains Peafields Legume hay Oilmeal 2-3 lbs.
Southwest	Range Prairie or alfalfa hay Cottonseed meal in winter 2-3 lbs.	Silage grain sorghum or corn 25-30 lbs. Cottonseed cake 5-6 lbs. Alfalfa or prairie hay 8-10 lbs.

It is possible in some districts of the blue-grass section to substitute peanut meal and velvet bean meal for cottonseed meal, especially on farms possessing suitable grinders for this purpose. In the southeast section the main reliance in growing steers must be placed on cottonseed meal and hulls; the velvet bean fields, the Bermuda grass pasture and the peanut vine hay supplementing in

PROGRESSIVE BEEF CATTLE RAISING

the periods when the cottonseed is not available. For fattening, the meal produced by grinding the entire velvet bean plant and the entire corn plant, gives a bulky and nutritious feed which the steer can use profitably up to the physical limits of its consumption. The northwest presents two distinct problems, the range problem and the small farm problem. The small farmer has a variety of products available for cattle growing, but the range producer is limited to the grass of his acres and prairie hay, silage or straw in the winter.

Growing the Calf

Success in growing cattle for market depends upon two things, the breeding of the calf and the start in life the calf receives. When calves are intended for straight beef production only a small quantity of feed in addition to milk is necessary up to weaning time, but they should be taught to eat supplemental feeds during this period to prevent a set back when milk no longer is furnished. The amount of dry feeds consumed will be limited at first but should be increased gradually until the calf no longer needs milk when six to eight months old. Calves intended for baby beef should begin on grain when four to six months old, a mixture of equal parts ground by weight of corn, oats, and wheat bran is good to start with, and after the calves have become accustomed to it, it may be fed whole. There is less danger of digestive disturbance and scours when the corn and oats are whole than when ground. The grain allowance should be increased gradually so that weaning time will not provide a set back to the calves. From then on calves intended for baby beef should be kept on full feed. The following rations may prove suitable to different sections of the country:

RATION No. 1

Corn..10 lbs.
Cottonseed Meal..2 lbs.

PROGRESSIVE BEEF CATTLE RAISING

Clover Hay.................................... 3 lbs.
Silage..12 lbs.

Oil meal and alfalfa may be substituted for cottonseed meal and clover, and calves may be allowed all the straw they will eat.

Ration No. 2
Corn.. 6 lbs.
Cottonseed Meal.............................. 3 lbs.
Legume Hay...................................10 lbs.
Straw......................................No limit

Ration No. 3
Kafir or Milo.................................12 lbs.
Cottonseed Meal............................. 2½ lbs.
Silage..12 lbs.

Ration No. 4
Barley or Broken Wheat........................10 lbs.
Roots...10 lbs.
Alfalfa Hay................................... 6 lbs.
Coarse Hay.................................No limit

In the foregoing rations oil meal may always be substituted for cottonseed meal, barley or kafir for corn, and any of the legume hays (alfalfa, clover, velvet bean, cowpea, soybean, lespedeza or peanut vine) for any other. It is important that the feed of the growing steer calf contain plenty of protein and mineral matter, hence such feeds as clover, alfalfa, and silage should be given in abundance with some oats, and cottonseed or linseed meal. Calves that are to be fed out as long yearlings or two-year-olds, or to be sold as stockers at a year old, may be fed quite largely the first winter on cheap roughages but it pays to give small amounts of concentrates in order to keep the calves growing in a thrifty condition.

It is highly important that calves be castrated when young, usually at six to eight weeks of age, because there is less danger of checking growth. The object of this is to prevent reproduction, to increase the fattening propensity, to make the animal easier to handle, and to improve the quality of the meat. One of the greatest dangers to livestock improvement comes from permitting calves with only one or two crosses of improved blood to grow into bulls, thus replacing well bred bulls on farms and ranches. In order to insure success in castration, one must carefully wash and disinfect the hands and instruments before operating and the wound after operation, must make a large free opening to permit good drainage and prevent pus accumulations in the wound and must permit the calf plenty of exercise to keep the swelling down. Calves turned to pasture immediately after the operation recover more quickly than those confined to the stable, since the chances of infection are less.

If calves are to be turned off as veals it is not necessary to castrate and they should be pushed along with skimmed milk, flax-seed meal and such other feeds as they can learn to consume, until they are six to twelve weeks old and fat enough to market. Great care must be taken to see that the skimmed milk is sweet and not fed in dirty receptacles, as the digestive system of the calf may be deranged and scours or some other ailment result. Very few calves of beef breeding are killed as veals, the majority of such calves coming from milking herds.

The Advantage of Young Cattle

There are certain general principles connected with the feeding of cattle that each farmer should bear in mind. The younger the animal the cheaper the gain.

The older the feeder the easier to fatten.

The older the cattle the greater the proportion of roughage consumed.

The older the cattle the less the labor and shelter required.

The greater the abundance of pasture and cheap feeds and the more limited the fattening feeds, the greater the profit in marketing cattle as stockers and feeders.

Older cattle digest their feed less closely than young cattle, and both digest whole grain less closely than ground grain.

The more limited the feeding space and the greater the supply of concentrates, the greater the opportunity for hogs to follow cattle.

One pig weighing seventy to eighty pounds should be allowed for three steers.

Essentials of a Complete Ration

After all the big problem in feeding is to satisfy the requirements of the animal, and a study of the requirements is necessary in order to feed scientifically and economically. Generally speaking the needs of the animal may be grouped under four heads: *Growth, energy, fattening,* and *health regulation*. *Growth* is dependent on the nitrogenous substances in the feed, and is supplied by such feeds as bran, milk, cottonseed meal, linseed meal, gluten feed, cowpeas, soybeans, alfalfa, and clover. Feeds which supply *energy* consist mainly of carbohydrates and fats, and are furnished by corn, barley, wheat, rye, prairie hay, straw, fodders, silage, grass, etc. *Fattening powers* are furnished by the same feeds as those that supply energy. At one time it was supposed that a simple estimate of the amounts of these feeds that would furnish a well balanced diet was sufficient in order to have successful results in feeding, but it is now known that there are certain ingredients of the ration that have only a slight food value, but that promote the utilization of the feed and the general *health* of the animal. Foremost among these may be mentioned mineral matter, such as salt, lime, etc., which is known to be essential to successful feeding, but there is another class of substances known as vitamines found in certain fats and proteins that promote the body processes in much the same way that lubricating oil promotes the work of the tractor without contributing to the energy that runs it. This is found in the hulls of cereals, cottonseed, flaxseed, timothy, some roots and alfalfa and clover hays. Corn is notoriously lacking in some of these substances, and animals fed corn alone get the "burned-out" appearance due to the lack of certain of these essential compounds in the feeds. Except for these health promoting substances however, one feed in a certain class may be substituted for another, according to cheapness in a given community, as bran for linseed meal or cottonseed meal, and kafir for corn or barley. Because of this possibility of substitution it is important that the feeder should study feed values, as only by such a system can he make a business-like profit on his feeding operations.

Silage

One of the most important feeds from the standpoint of health promotion is silage, which provides the steer and growing animal with succulent green feed the year around. While corn is pre-eminently the best silage plant, kafir, sunflowers, cane, oats and peas, alfalfa, and soybeans with oats make a very desirable product. The principle involved in ensiling feeds is manifold: in general fermentations change the sugars of the plant to acids, and preserve the total food value more perfectly than any other method of feed preservation. If there is a shortage of starches and sugars as in alfalfa, peas and beans, it must be made up by mixing corn, cane, oats, rye, or some similar agent. Molasses has been used satisfactorily where cheap enough, being sprinkled over the green legume as it goes through the silage cutter. The two most important provisions in silage making are the presence of these sugars and the exclusion of air. If too much air is present, the silage putrefies, hence one must be careful to tramp such crops as oats very carefully in order to drive the air from the hollow stems. There are six distinct advantages from ensiling crops. The relative expense is low ($2.50

to $7.50 a ton, depending on investment in silo and machinery and labor conditions), it can be made available for any season of the year, less of the feed value is wasted, it is eaten with practically no waste, the weather handicaps the making of silage less than putting the crop up in any other form, it makes weeds available for feed if mixed with the silage crop, and it can be stored in less space than the same feed dry, in the ratio of 2 to 5 as far as food value is concerned.

Requisites of a Good Silo

The requisites of a good silo are: 1—airtight walls; 2—cylindrical shape (to prevent corners which fill improperly); 3—smooth, strong, perpendicular walls (to prevent air pockets); and 4—depth (to give pressure on the mass of fermentling feed, to reduce the percentage loss through fermentation of top layers before they can be fed, and to reduce the loss of food nutrients, which are greatest in upper part). Silos may be made of staves, brick, hollow tile, concrete, stone or steel. Pit silos with cement lining and concrete curb may be used in arid and semi-arid climates, but the material used anywhere for structure depends upon local conditions. See illustration on page 32.

Silo Capacities

The diameter of a silo to be erected should be determined from the number of animals to be fed, the idea being to feed about two inches of silage off the top to prevent spoilage. The minimum amount to be fed daily, to attain this depth of feeding, is shown in the following table, allowing twenty-five pounds per head:

Diameter of Silo	Minimum Amount of Silage	Number Head
10 feet	520 pounds	21
11 feet	625 pounds	25
12 feet	745 pounds	30
14 feet	1,015 pounds	41
16 feet	1,325 pounds	53

Diameter of Silo	Minimum Amount of Silage	Number Head
18 feet	1,680 pounds	67
20 feet	2,075 pounds	83
22 feet	2,510 pounds	100
24 feet	2,985 pounds	119
26 feet	3,505 pounds	140

In order to determine the capacity of a silo, multiply the number of cubic feet in the silo by 40 pounds, the average weight of a cubic foot of silage. As a matter of fact, this figure will vary according to the height of the silage in the silo, and those interested in a really accurate result can use the following average weights:

Depth of Silage	Average Weight for Whole Depth	Weight at Given Depth
1-foot	18.7	18.7
10-feet	26.1	33.1
20-feet	33.3	46.2
30-feet	39.6	56.4
36-feet	42.8	61.0

The deeper the silage gets the more a cubic foot of silage at that depth weighs and the heavier the average of all the silage is.

Part V.

Management of the Beef Herd

Three Types of Cattle Farming

There are three systems of handling beef-bred herds in common usage in the United States. The *straight beef system* in which the steers are grown out as cheaply as possible is adapted to regions where pasture is plentiful and cheap and is practiced more widely in United States than any other method of beef production. The *dual purpose system* is used more commonly in the general farming states although up to the present it is not more popular than the straight beef system if the numbers practicing it be any criterion. In this system the cows are milked and the calves are raised on skimmed milk and supplemental feeds. The dual purpose calves as a rule are not as economical beef producers as the straight beef calves but when grown out and fattened they frequently make very acceptable beef. The dual purpose system is commendable only when adhered to properly, and is likely to be quite unsuccessful if it is attempted to turn the beef animals into a dairy herd. The *baby beef system* is a highly specialized method and is adapted to such districts as the cornbelt where there is a good supply of feeds for fattening and sufficient pasture for the summer maintenance of the breeding cows with their calves. While it requires a little more equipment to handle the herd the best market prices can be obtained in baby beef as well as in the dual purpose systems, if the calves are dropped in the fall and finished to market in the summer and early fall. If calves are dropped in the spring they should come late in February, March or early April, but if they come in the fall, late August, September and early October are preferable. The question as to the better time can be settled only by a study of individual farm conditions, taking into consideration the equipment, labor, pasture and feed supply.

The Maintenance of the Breeding Herd

Cows raised for the production of calves only, can be fed very cheaply during the biggest portion of the year by using silage and dry roughages combined with a small quantity of such feed rich in protein as oilmeal or cottonseed meal. If clover or alfalfa hay is available, these may be omitted except during the periods immediately following calving and for two weeks before breeding. Such cows do not require anything more than open shelter except at calving time, when they must be placed separate from the rest of the herd. If fall calving is practiced little shelter for the cow at parturition is required, but if the calves come in February, March and April, both dam and offspring must be sheltered from the extremes that sometimes occur at that season of the year. The purchase of feeds for breeding cows should not be discouraged when necessary, since a suitable purchase may be more than repaid in the additional growth of the calf. Successful cattle raisers must grow the necessary roughages however, and for this part of the ration can well adopt the slogan "Grow all you feed and feed all you grow." In the summer the cow herd will be maintained largely on pasture but if the pastures are short supplements must be provided. Silage is the best agent for this but if not available dry roughage such as hay or green forage crops should be provided. After harvest, the cows can be maintained for a time on the stubble and grass growth in the fields, in fact some men plant clover or other crops which will develop after harvest for this very purpose. In the South, velvet beans may be utilized for the pasture of fall and early winter while farther north the stalk fields are available. In the winter hay and silage will provide the main dependence but when protein feeds are nec-

PROGRESSIVE BEEF CATTLE RAISING

essary from one to two pounds of linseed or cottonseed cake may be fed.

The Pasture The fundamental requirement for economical beef cattle production is plentiful and permanent pasture. In the cornbelt, bluegrass pasture has proved to be the most satisfactory permanent proposition, but white clover mixtures make a little richer feed of it. From a temporary standpoint good returns may be obtained from mixed timothy and red clover while in shaded areas orchard grass and red top should be used. In the range country native grasses have been found superior to anything seeded, but care must be taken not to overstock them or both variety and amount of herbage are lost. Considerable success has been found in parts of Kansas and Oklahoma by restricting the pasturage on certain areas and seeding the remainder in order to renew the growth. The farmer who feels that his pasture is deserving of a little investment and care will find the distribution of suitable fertilizers will promote the growth of grass very decidedly. Advice as to the kind of fertilizer or the amounts should be obtained from the state experiment station or from the county agricultural agent. In the South, Bermuda grass and lespedeza have been found highly resistant to drouth and their use in southwestern states may possibly be extended. The farmer must remember that while the grasses are natural in most of the sections of America, they are not spontaneous under heavy systems of pasturage, and discing, seeding, and occasional fertilizing are necessary to obtain the greatest returns.

The Contents of the Hay Stack Hay provides the winter substitute for pasturage on most farms. The successful farmer will calculate the amount of hay he needs to carry his cattle through the winter, allowing from ten to sixteen pounds per head, per day, depending on the availability of

PROGRESSIVE BEEF CATTLE RAISING

such other feeds as straw, corn stalks, and silage. In order to calculate this it is desirable for him to know how to determine the number of tons of hay in the stack. The ordinary method of determining this is first to find the volume of the stack in cubic feet and then to

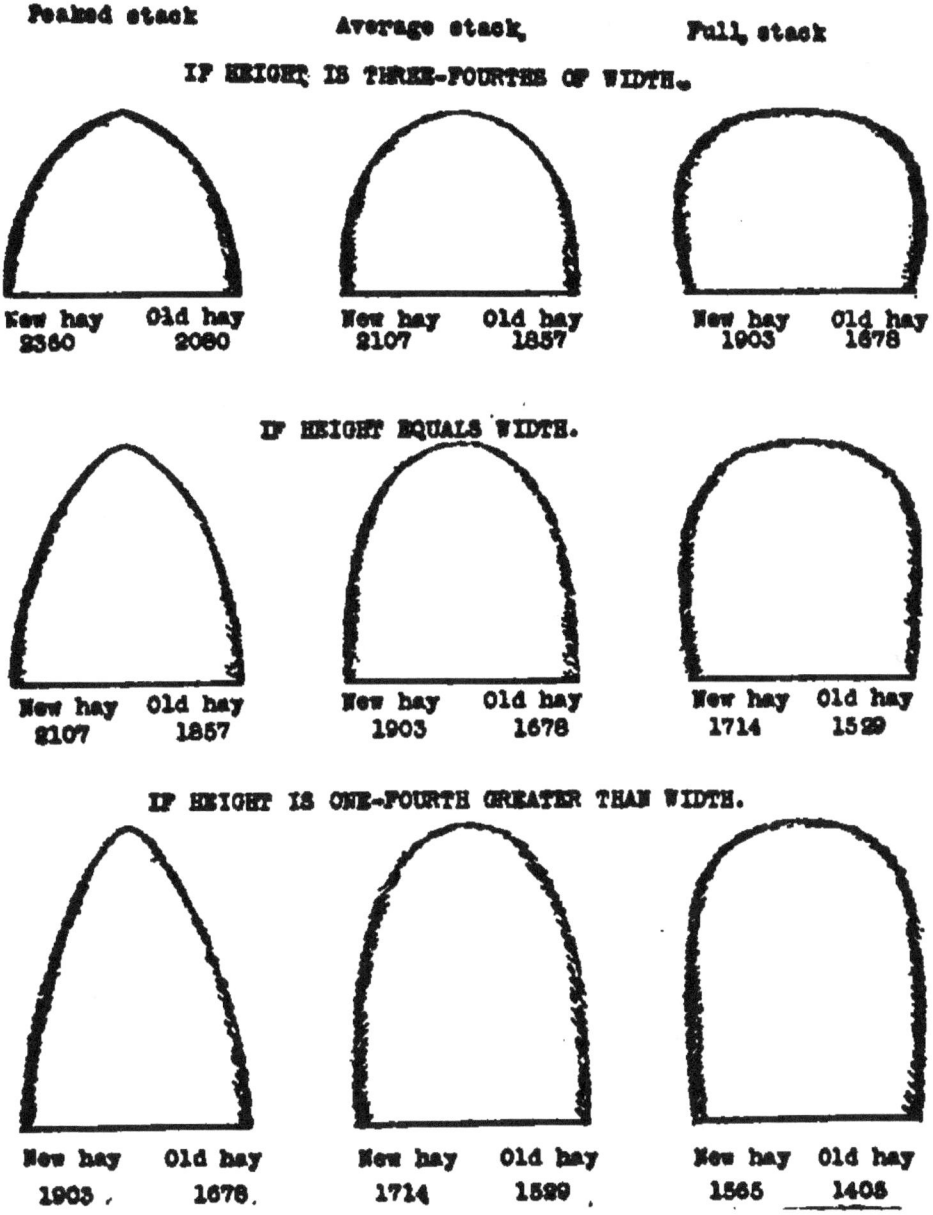

To determine contents of hay stack multiply length by width by over and divide by the number indicated above in order to obtain the number of tons.

transform it to tons. To determine this the farmer will measure the width and length of the stack and then get the distance from the ground on one side to the ground on the other at a point which is about the average height of the stack. Having obtained these three figures for width, length and over, they are multiplied together and divided by the figures shown in the accompanying diagram depending on the shape of the stack and the length of time the hay has been in the stack. The resultant figure will give the number of tons of hay in the stack.

Sanitation on the Farm

One of the most important factors in success with beef cattle is the health of the herd. The cheapest way in the long run to safeguard the breeding animals is by the prevention of disease and sanitation. Every cattleman should provide himself with an isolation shed and pen to which sick animals can be taken. This will secure privacy and rest for the animals and in addition will limit the spread of contagious or infectious disease. After a diseased animal has been removed from this lot all straw and manure not exposed to the sunlight and wind should be carefully burned and the shed and feed troughs should be disinfected, either with lime or a spray. Special care should be taken to provide bright, clean, sanitary quarters for calving as a step taken in time here may prevent serious losses later due to hemorrhagic septicemia, scours, sore eyes, snuffles, and various other calf diseases. Feed troughs, water tanks, and other places where cattle commonly come in contact with each other should be kept scrupulously clean and should be disinfected frequently.

Cattle Diseases Some of the commonest diseases which American cattlemen have to face are lump jaw, blackleg, contagious abortion, foot and mouth disease, foot rot, hemorrhagic septicemia and tuberculosis, while the following are the commonest parasites which have to be combated; Texas fever tick, lice, screw worms, ox warble and mange.

LUMP JAW. Lump jaw is a chronic non-infectious disease that affects the jaws of cattle and the udders of swine. It is caused by a fungus that is frequently found on barley beards, oat stubble and various grasses, although it does not grow outside of the animal body. It appears as a hard tumor-like swelling on the jaw in the early stages of the disease, but later becomes ulcerated from the inside, causing slobbering and difficulty in chewing. The animal becomes emaciated and frequently starves to death. The most satisfactory way to handle the animal is to begin fattening it at the first signs of disease and ship to market before the affection becomes too marked. Such animals are subjected to rigid examination after death and if the disease is localized in the head the animal is passed as fit for food.

BLACKLEG. This is a highly contagious disease that affects cattle between the ages of six and twenty-four months. It is usually fatal in the course of twelve to thirty-six hours after the animal first shows signs of sickness. However, the animal may have been infected from three to five days previous to the first symptoms. The animal shows a high fever, loss of appetite and great depression, while it usually stops chewing its cud in the very earliest stages. Swellings appear over the heavily muscled parts of the body and if one strokes the skin in these parts a distinct crackling is heard and felt. There is no satisfactory remedy and the best method is preventive treatment by means of vaccination. The losses following this are less than one-half of 1 percent.

CONTAGIOUS ABORTION. This is a chronic and highly insidious disease that is confined to the organs of reproduction and is probably the most widely spread disease in cattle. It is caused by a specific germ which is more likely to infect heifers than cows and which seldom affects the bull. During the early months after breeding the animal appears normal but the calf may be born from three to five months prematurely. Some cows may become "carriers" of the disease without themselves being sick. Skilled veterinarians are required to recognize these animals by means of blood tests. The only treatment possible is preventive. Immediately after the animal aborts all of the litter should be disposed of by burning and the stable floor should be disinfected with a strong liquid. The cow should be douched with a 1 percent solution of salt at blood temperature to prevent the accumulation of pus. Some investigators at present urge the use of a vaccine, but this has not yet been perfected.

FOOT AND MOUTH DISEASE. Although this disease is not common in America, there have been several serious scourges from it, the last in the years 1914-15. It affects cattle worse than other stock and the mortality ranges from 1 to 3 percent. The disease opens with a moderate fever and the appearance of blisters in the mouth and between the hoofs. A profuse flow of saliva is stimulated which hangs from the mouth in viscid ropes. No attempt is made to treat the disease in the United States and infected animals are immediately slaughtered.

FOOT ROT. Cattle that are forced to stand in filthy lots occasionally suffer from a contagious hoof disease known as foot rot. The animals become lame, develop a hot and painful swelling around the hoof and lose their appetite and flesh. The "proud flesh" which appears must be trimmed away, the pus tracts drained and a disinfectant applied. In cattle the best remedy is pine tar held in place by a bandage passed between the claws and tied around the pastern.

HEMORRHAGIC SEPTICEMIA. This disease runs a short course in cattle that frequently ends in death and affects calves more commonly than older animals. The method of infection is not known although cattle on pasture are less likely to be affected than those under confinement. The animals refuse feed, exhibit a severe fever, show difficulty in breathing and develop swellings in the throat and brisket. When the intestines are affected the animals show signs of colic and pass bloody manure. Once the disease is developed medicines are useless, hence efforts are directed toward preventing the spread to other animals. All unaffected animals should be removed to fresh quarters and vaccinated, and the infected buildings and lots disinfected.

TUBERCULOSIS. This is one of the most serious diseases affecting cattle because of the possibility of its transmission to man. It is readily transmitted to hogs following cattle, in many cases as high as 25 percent being rendered unfit for food. The disease is so named because small tubercles form in the internal organs. Infection is usually spread by eating food or drinking fluids contaminated by the discharges from infected animals. Frequently animals severely afflicted with the disease show no signs of it externally. If the lungs are affected there may be a cough and difficulty in breathing, while if the intestines are involved, a chronic diarrhea is present. The two common tests for the presence of tuberculosis are the injection test and the intradermal test with tuberculin. In the first case the animal shows a marked rise in temperature a few hours after injection if affected with the disease, while in the second case a small hard swelling develops at the point of inoculation within twenty-four to ninety-six hours. Treatment is unsatis-

PROGRESSIVE BEEF CATTLE RAISING

factory and the only practicable method known is the preventive one which removes all infected animals and utilizes sanitary methods.

TEXAS FEVER TICK. Only a few years ago this pest was prevalent throughout the southern states, but is being rapidly eradicated by means of the quarantine. Its ill effects come through the injury to the cattle in sucking their blood and infecting them with the germs of a disease that results in high fever and occasionally death. The vitality of most infected animals is so low that they are not profitable to handle. The most successful means of getting rid of the tick is by periodic dipping.

CATTLE LICE. These parasites do the most damage in the winter months and are more likely to infect thin cattle than fleshy ones. They can best be disposed of by dipping in the fall before cold weather sets in, followed by a second dipping seven to ten days later to kill any lice hatching after the first treatment.

SCREW WORMS. During hot weather screw worms may appear in wounds, cuts or sores, as a result of eggs laid in these parts by a fly. The most effective treament is to open the wounds, to wash them with gasoline, and to daub them with pine tar.

WARBLES. The ox warble is a grub which develops under the skin in late winter or early spring, bores a hole through it, and drops to the ground where it hatches into a fly. There are no preventive measures known but the grubs ready to drop from the animal should be squeezed out and destroyed and those not quite ready to emerge should be dislodged with a sharp knife.

MANGE. Mange is caused by a small mite that attacks the skin causing it to become scurfy. It spreads from one animal to another by contact and can be remedied by dipping or spraying.

Cattle Diseases

Some of the commonest diseases which American cattlemen have to face are lump jaw, blackleg, contagious abortion, foot and mouth disease, foot rot, hemorrhagic septicemia and tuberculosis, while the following are the commonest parasites which have to be combated; Texas fever tick, lice, screw worms, ox warble and mange.

LUMP JAW. Lump jaw is a chronic non-infectious disease that affects the jaws of cattle and the udders of swine. It is caused by a fungus that is frequently found on barley beards, oat stubble and various grasses, although it does not grow outside of the animal body. It appears as a hard tumor-like swelling on the jaw in the early stages of the disease, but later becomes ulcerated from the inside, causing slobbering and difficulty in chewing. The animal becomes emaciated and frequently starves to death. The most satisfactory way to handle the animal is to begin fattening it at the first signs of disease and ship to market before the affection becomes too marked. Such animals are subjected to rigid examination after death and if the disease is localized in the head the animal is passed as fit for food.

BLACKLEG. This is a highly contagious disease that affects cattle between the ages of six and twenty-four months. It is usually fatal in the course of twelve to thirty-six hours after the animal first shows signs of sickness. However, the animal may have been infected from three to five days previous to the first symptoms. The animal shows a high fever, loss of appetite and great depression, while it usually stops chewing its cud in the very earliest stages. Swellings appear over the heavily muscled parts of the body and if one strokes the skin in these parts a distinct crackling is heard and felt. There is no satisfactory remedy and the best method is preventive treatment by means of vaccination. The losses following this are less than one-half of 1 percent.

CONTAGIOUS ABORTION. This is a chronic and highly insidious disease that is confined to the organs of reproduction and is probably the most widely spread disease in cattle. It is caused by a specific germ which is more likely to infect heifers than cows and which seldom affects the bull. During the early months after breeding the animal appears normal but the calf may be born from three to five months prematurely. Some cows may become "carriers" of the disease without themselves being sick. Skilled veterinarians are required to recognize these animals by means of blood tests. The only treatment possible is preventive. Immediately after the animal aborts all of the litter should be disposed of by burning and the stable floor should be disinfected with a strong liquid. The cow should be douched with a 1 percent solution of salt at blood temperature to prevent the accumulation of pus. Some investigators at present urge the use of a vaccine, but this has not yet been perfected.

FOOT AND MOUTH DISEASE. Although this disease is not common in America, there have been several serious scourges from it, the last in the years 1914-15. It affects cattle worse than other stock and the mortality ranges from 1 to 3 percent. The disease opens with a moderate fever and the appearance of blisters in the mouth and between the hoofs. A profuse flow of saliva is stimulated which hangs from the mouth in viscid ropes. No attempt is made to treat the disease in the United States and infected animals are immediately slaughtered.

FOOT ROT. Cattle that are forced to stand in filthy lots occasionally suffer from a contagious hoof disease known as foot rot. The animals become lame, develop a hot and painful swelling around the hoof and lose their appetite and flesh. The "proud flesh" which appears must be trimmed away, the pus tracts drained and a dis-

RING WORM. This disease is quite similar to mange but causes circular patches on the skin instead of a general infection. It is most common during the winter and spring and is usually found on the heads and necks, although it may affect any part of the body. It causes severe itching and is remedied with iodine and nitrate of mercury ointment. Stables should be disinfected.

The Cow and Her Calf

As a general practice it is advisable to provide quarters for calving even though it may not be necessary to use them ordinarily. The average breeding cow needs little assistance if she is in a vigorous, healthy condition, nor do most calves, but there are many that die which would have lived if assistance had been available at the proper time. As soon as the calf is born all membranes should be removed from the mouth and nose and if the calf is not strong, a slight pull on the tongue and pressure in the ribs may stimulate breathing. The cow should be allowed to dry the calf herself and to give it its first care, although the calf may need assistance the first time to find the udder. The calf should always receive the first milk from the udder unless the cow is feverish and her udder inflamed, since it acts as a mild purgative. Clean, sanitary quarters are a distinct asset to any breeding farm.

Gestation Table

For convenience in determining the time the cow is due to calve, the time of service being known, a gestation table is given on page 42, by the use of which it is very easy to determine the approximate time a cow will calve. It will assist in keeping accurate breeding records.

PROGRESSIVE BEEF CATTLE RAISING

Gestation Table for Cows (283 Days)

EXPLANATION: Find date cow was bred in first column and month bred in top line. The date in column below opposite date bred will be the time at which the cow is due to calve.

Day of Mo'th Bred	Jan. / Oct.	Feb. / Nov.	Mar. / Dec.	Apr. / Jan.	May / Feb.	June / Mar.	July / Apr.	Aug. / May	Sept. / June	Oct. / July	Nov. / Aug.	Dec. / Sept.
1	11	11	9	9	8	11	10	11	11	11	11	10
2	12	12	10	10	9	12	11	12	12	12	12	11
3	13	13	11	11	10	13	12	13	13	13	13	12
4	14	14	12	12	11	14	13	14	14	14	14	13
5	15	15	13	13	12	15	14	15	15	15	15	14
6	16	16	14	14	13	16	15	16	16	16	16	15
7	17	17	15	15	14	17	16	17	17	17	17	16
8	18	18	16	16	15	18	17	18	18	18	18	17
9	19	19	17	17	16	19	18	19	19	19	19	18
10	20	20	18	18	17	20	19	20	20	20	20	19
11	21	21	19	19	18	21	20	21	21	21	21	20
12	22	22	20	20	19	22	21	22	22	22	22	21
13	23	23	21	21	20	23	22	23	23	23	23	22
14	24	24	22	22	21	24	23	24	24	24	24	23
15	25	25	23	23	22	25	24	25	25	25	25	24
16	26	26	24	24	23	26	25	26	26	26	26	25
17	27	27	25	25	24	27	26	27	27	27	27	26
18	28	28	26	26	25	28	27	28	28	28	28	27
19	29	29	27	27	26	29	28	29	29	29	29	28
20	30	30	28	28	27	30	29	30	30	30	30	29
21	31	Dec. 1	29	29	28	31	30	31	July 1	31	31	30
22	Nov. 1	2	30	30	Mar. 1	Apr. 1	May 1	June 1	2	Aug. 1	Sept. 1	Oct. 1
23	2	3	31	31	2	2	2	2	3	2	2	2
24	3	4	Jan. 1	Feb. 1	3	3	3	3	4	3	3	3
25	4	5	2	2	4	4	4	4	5	4	4	4
26	5	6	3	3	5	5	5	5	6	5	5	5
27	6	7	4	4	6	6	6	6	7	6	6	6
28	7	8	5	5	7	7	7	7	8	7	7	7
29	8	.....	6	6	8	8	8	8	9	8	8	8
30	9	.....	7	7	9	9	9	9	10	9	9	9
31	10	.....	8	.....	10	.....	10	10	.....	10	.....	10

PART VI

The Cattle Industry

The United States Position in Beef Production

As a producer of beef the United States leads the world. On January 1, 1920, there were 68,132,000 cattle reported by the Bureau of Crop Estimates, of which 44,385,000 were beef animals. These exceeded the nearest competitive beef country by over a third. For 1919 the relative figures on cattle, including milk and draft animals as well as beef, were given as follows by the Year Book of Figures of the Chicago Daily Drovers' Journal:

India	147,335,000*
United States	67,866,000
Russia	52,052,000
Brazil	28,962,000
Argentina	25,867,000
Germany	20,317,000
France	12,189,000
Canada	10,051,000
Australia	9,924,000
Uruguay	8,193,000

Not all of these countries are meat surplus countries, however, as many of them consume more than they produce. The nine principal meat export countries are the United States, Argentina, Australia, Canada, New Zealand, Uruguay, Mexico, Denmark and Paraguay. Brazil is rapidly progressing in beef cattle production and the day is not far off when she, too, will be a factor in the world's beef markets.

---

*Includes buffaloes and millions of draft cattle. Proportion of beef cattle relatively low.

PROGRESSIVE BEEF CATTLE RAISING

The American Beef Export Trade

During the war America exported large quantities of beef which permitted high prices to the producer. During the five years 1910-1914 the average exports of beef products totaled 71,717,597 pounds, while during the five years of war, 1915-1919, the average exports were 385,612,976 pounds, an increase of approximately 538 percent. America's principal customers for the last seven years are indicated by the following table:

Country	1913	1914	1915	1916
United Kingdom	26,292,537	19,353,985	152,865,431	199,583,322
France	153,430	67,542	106,455,420	60,162,930
Italy	408,775	437,556	11,871,446	53,633,048
Netherlands	50,661,952	48,888,764	37,183,392	30,373,211
Scandinavia	15,850,056	15,677,189	34,389,858	36,275,700
Germany	23,809,562	19,572,376	1,393,312	850

Country	1917	1918	1919
United Kingdom	205,616,173	390,033,935	347,932,906
France	58,672,952	68,800,838	53,152,973
Italy	14,019,782	17,634,774	65,045,368
Netherlands	14,040,591	....	329,694
Scandinavia	24,825,785	1,417,317	22,481,954
Germany	....	....	....

Relation of Export Trade to Cattle Production

During times of peace the export trade is a very minor factor in supporting beef prices. In the past ten years we have exported the following percentage of our annual beef crop:

	Total Beef Production	Total Beef Exports	Percent Beef Exported
1910	9,621,000,000	127,405,575	1.3+
1911	9,398,000,000	93,618,984	1.0—
1912	8,997,000,000	64,378,658	0.7+
1913	8,559,000,000	40,059,655	0.5—
1914	8,236,000,000	33,125,114	0.4+
1915	8,728,000,000	277,558,938	3.2—
1916	10,224,000,000	320,132,447	3.1+
1917	12,779,000,000	322,766,893	2.5+
1918	10,200,000,000	521,844,093	5.2—
1919	9,050,000,000	485,762,509	5.4—

PROGRESSIVE BEEF CATTLE RAISING

The war increased the percentage of our beef exports almost five times, the average for the first five years quoted above being .8 percent, while for the last five years it was 3.9 percent. The figures quoted above are for the fiscal years closing June 30. They do not include the period of the great drop in exports which occurred during the last half of 1919. If one compares the calendar years 1918 and 1919, one factor in the dropping beef prices of that period becomes apparent, for exports of 699,996,712 pounds in 1918 drop to 346,347,892 pounds in 1919. It is difficult to say just where the percentage of exports becomes large enough to be important, economic and political conditions having greater effect perhaps than any mathematical relation.

Part VII

Cattle Prices

The Relation of the Market to the Feeding Business

The market has a very intimate relation to beef cattle production. The effects are not immediate but are reflected from four to twelve months later. High prices bring high receipts at the cattle markets, but the producer saves enough females to enlarge his marketing possibilities for the next year or two. On the other hand, lowering prices tend to fall still lower because the producer sees no hope ahead for expansion and turns females that should be breeding onto a market already depressed. The most successful feeders have had the best results by going just contrary to inclination of the average cattleman, since conditions are nearly always reversed by the time the next crop of animals can be made marketable. Such a feeder makes cattle the medium for marketing certain portions of the farm's rough products yearly and thereby makes feeding a permanent business. When returns from such a system are considered over a period of years it will be found that the losses of one year are absorbed by the profits of another, with a reasonable margin for the feeder, while his land will have been permanently upbuilded by the system. Beef cattle feeding is operated on as narrow a margin as obtains in any farm operation and very slight fluctuations in prices may reduce profits to losses or vice versa. Since such fluctuations exist it is very easy for men to lose badly or to make large gains, but since they are difficult to foresee, there are very few cattle speculators except those close to market that have been financially successful. The only real winners in the beef cattle business are those who have made it an integral part of their farming operation, year in and year out.

PROGRESSIVE BEEF CATTLE RAISING

Why Markets Fluctuate

Price fluctuations provide the sorest spots in marketing that the producer faces. It has never seemed logical to him that a steer worth one price today may be worth a different price a week from today when the costs of the feeds and labor as he has handled them, have not varied. Furthermore the food value of the meat from the steer is approximately the same during a period of several weeks in the finishing of beef animals. From the producer's viewpoint these ideas seem perfectly proper, but producers should take into account the fact that *the main factor in meat prices is the existence of a ready market*. Farm produce, except fruits and vegetables, does not fluctuate like finished beef products, because it is not quickly perishable. Once it is put into a perishable form, however, farmers and feeders must take chances with others. Beef is just such a perishable article and unless the market can move out of the coolers and refrigerators a sufficiently steady stream of meat to make place for that incoming from the daily kill of cattle, slaughter at the prices being currently paid becomes almost impossible. Lowered prices do two things, they give a slight margin which will help pay the charges on such beef as has to be stored for a few days, and they cause the producer to hold up shipments until a more favorable price may be received, thereby permitting the market to work off excess supplies. As long as the consumer varies in the amount of meat he eats daily, thereby affecting the buying power of the retailer, just so long will there be sudden fluctuations in the prices of beef.

The Two Classes of Price Fluctuations

There are two general classes of price fluctuations. One class is expressed as trends, due to causes that affect prices over a considerable period of time, while the other is the day to day fluctuation that results from the variations in receipts on the market, variations in the ability of the markets to move the beef on hand, and the competition of other meats and foods for the favor of the family pocketbook. Illustrations of factors causing general trends in prices are the seasonal changes in the meat appetite of the buying public, the seasonal variations in receipts, the influence of the export trade, and such unusual occurrences as the recent war. *The most important factor of all of these is the ability to sell beef and this is almost perfectly correlated with the volume of beef on hand as related to volume of business, and is entirely unrelated to the cost of production of the cattle.*

Seasonal Variations in Price

One of the chief factors contributing to low prices for the average feeder is the tendency for every cattleman to market his steers in the period November to April. The chart opposite presents a study of the prices of native beef on the Chicago market over a period of twenty years and shows that prices for this class of cattle are above the average of the year from April to mid-October, and below the average for the remainder of the year. Of course there have been years in which this was not true, but it represents the average condition over this time. The deviations below the average price for the year are greatest in January, February, June, November and December, while the least occurs in the period July to October. Furthermore, in the November to May period, the monthly average price runs nearer the bottom of the deviations from the average annual price, while in the period May to October, the monthly average runs nearer the top of the deviations from the average annual price. This means that the man who markets in the latter period not only gets better prices for his stock, but that he is much more likely to top the market, simply because the monthly average tends to run nearer there.

PROGRESSIVE BEEF CATTLE RAISING

The Problem of Marketing Beef in All Seasons
Because prices are higher at this time does not necessarily mean that the farmer feeder will get greater profits. Beef production depends on the annual cycle of the year, steers fatten on the crops that mature during the year, but meat is eaten on the daily cycle based on the frequency with which man gets hungry. It therefore happens that the bulk of the cattle come on the market in the months September to February, after they have eaten the crops of the preceding season. Man's appetite runs throughout the year, however, and if he is to be assured a supply of meat at all times, someone must carry the cost of holding either animals or meat over to the leaner months. If the animal is killed the market must absorb the cost of handling or storing; if it is held over for better prices the feeder must pay this cost in feed, equipment and maintenance. It therefore becomes an individual problem for each farmer to determine, whether the additional costs of carrying his cattle over will be more than the rise in price he will probably get. Few are competent to advise with him except the men of his locality who are familiar with conditions. Possibly the county agricultural agent can be of service in this connection.

Methods of Reaching the Most Favorable Markets
The two simplest ways of taking advantage of the better markets in the seasons of lean supply are, first, to buy half finished cattle that some other shipper has put on the market and feed them for thirty to sixty days, depending on condition, and, second, to handle the animals by cheap maintenance and suitable farm equipment to carry over into this favorable time. The first system is best adapted to feeders who live close to a big livestock market. At almost any time during the winter the careful buyer can go on the market and watch for steers selling at prices he can afford to consider for further

PROGRESSIVE BEEF CATTLE RAISING

feeding. On the stockyards markets he will have to compete for warmed-up cattle with packer and speculator buyers to a greater extent than for the customary type of feeders, but he will find certain days in every week (not always the same day by any means), when there are more of this type on the market than can be absorbed and he can buy right. He may find it to his advantage to do this twice in a winter rather than once, in order to get the most economical use of his feed. He will need dry bottomed feedlots, and protection from the north and west in the more severe sections of the country, and he cannot profitably depend on the steers doing much rustling of their own, but must have facilities for bringing the steers' feed to them. The other system depends on finishing the steers on grass while it is still lush and requires silage and in some sections, soiling crops (peas, oats, vetch, cane, rye, etc.) as supplements. It adapts itself to baby beef production as calves of one spring can be marketed in July or August of the following year. Sufficient winter shelter will be required for these calves to keep them from using too much of their feed to provide heat. This system of feeding and finishing will do well on high priced land, but is of little value in the range and semi-range states where the system of production is too extensive.

The Effect of Supply and Demand on Hoof Prices

The relative effect of supply and demand on cattle prices on the hoof is shown in the chart on page 49. The two upper curves are not on the same scale, one square on the curve for market receipts representing 50,000 animals while one square for Armour's purchases represents only 4,000 animals. To make the curves directly comparable the heights in the first curve should be multiplied 12.5 times. In the case of the two lower curves, however, the general parallelism is quite marked. The fluctuation of hoof prices is not as great as that in dressed beef, because the price of byproducts does not vary with the price of beef but remains relatively constant. On the other hand it will be observed that the purchases of Armour and Company bear only an indirect relation to the price of dressed beef. In general, purchases were low when prices were high and vice versa, but there are almost as many exceptions as illustrations. This is simply another way of saying that prices went up when Armour and Company did not have the beef to supply, while with increased supplies, prices dropped.

If the points along the dressed beef curve and the live cattle curve are compared for divergence in direction, periods in which supply overrode demand to a slight degree will be noted. In September, 1917, and September, 1918, dressed beef prices rose while hoof prices dropped, but it will be noted that in each case the number of cattle on the market materially increased—in the first instance 65,000 head or 27.6 percent of the previous month's receipts, and in the second instance, 155,000 head or 64.6 percent. On the other hand, in November, 1918, and July and August, 1919, dressed beef prices dropped while hoof prices rose. This was unquestionably due to the fact that our purchases of the months preceding each of these periods were extremely light, while each preceded a period when the beef trade normally picks up, in the first case for Christmas, and in the second for the advent of cooler weather.

PROGRESSIVE BEEF CATTLE RAISING
===

PART VIII

The Beef Carcass

The Relative Value of Carcass Cuts There are eight standard wholesale cuts from the carcass; the round, the loin, the flank, the rib, the chuck, the plate and the shank, as shown in the illustration on page 56, and the suet secured from the free fat of the animal. There is a pronounced difference in the value of different carcasses and in the value of the cuts produced from different parts of the same carcass. The quality of the carcass is dependent on the relative thickness of the lean meat, its tenderness, the interspersion of fat among the muscle fibers, the firmness of the flesh, the freedom from bruised spots, the rich redness of the lean meat, and the clear white of the sound firm fat. Carcasses poorly protected by fat cannot stand handling in the fresh meat trade, while carcasses too darkly red in the lean and too yellow in the fat indicate age or finish on feeds that produce a more perishable carcass. Two very important factors affecting the value of the carcass are the lightness of the bone and the relative proportion of the valuable cuts. When beeves are handled in bulk, as in Armour and Company's Dressed Beef Department, the average proportion of the different cuts is usually figured for convenience in pricing, but if the carcasses are sold to the retailer, the proportion of valuable cuts is especially important. The demand as reflected from the retailer is shown by the different price per pound in the following table, in which the carcasses are considered by the cwt., thereby showing the percentages of the different cuts. The same price is allowed for the cuts from both carcasses in order to show the effect of the more valuable portions on the profits. In practice carcasses showing as great differences as recorded here would sell for different prices per pound.

PROGRESSIVE BEEF CATTLE RAISING

		Steer No. 1		Steer No. 2	
Cuts—Price per lb.		Wt. cut cwt.	Value	Wt. cut cwt.	Value
Round	$0.22	24	$5.28	22	$4.84
Loin	.34	18	6.12	16	5.44
Flank	.11	3.5	.385	4	.44
Suet	.145	3.5	.5075	4	.44
Rib	.27	10	2.70	8	2.16
Chuck	.12	25	3.00	27	3.24
Plate	.10	13	1.30	14	1.40
Shank	.095	3	.285	5	.475
Total			$19.5775		$18.575

In other words, on each hundred pounds bought in the proportions listed above, steers like No. 1, would be worth $1.00 more than steers like No. 2, even though their meat was of exactly similar grade. On a 650-pound carcass, this difference would be $6.50.

Factors in Carcass Values It is very seldom that two steers showing the difference in proportion of cuts cited in the foregoing would produce meat of similar value, but one carcass would be of lesser quality than the other. During the winter 1919-20 it frequently happened that the highest quality beef sold around 23 cents a pound wholesale, while good quality stuff was bringing about 18.5 cents. Two carcasses cutting up similar to those discussed in the foregoing paragraph would yield wholesale cuts as follows:

	Steer No. 1			Steer No. 2		
Cut	Percent	Price per lb.	Value	Percent	Price per lb.	Value
Round	24	$0.20	$4.80	22	$0.18	$3.96
Loin	18	.46	8.28	16	.39	6.24
Flank	3.5	.08	.28	4	.08	.32
Suet	3.5	.16	.56	4	.16	.64
Rib	10	.36	3.60	8	.30	2.40
Chuck	25	.16	4.00	27	.12	3.24
Plate	13	.10	1.30	14	.09	1.26
Shank	3	.07	.21	5	.07	.35
Total			$23.03			$18.41

PROGRESSIVE BEEF CATTLE RAISING

In this case two differences of importance exist between the two steers, *percentage of valuable cuts* and *quality of cuts*. The difference is expressed as $4.62 per hundred, or $27.72 on a 600-pound carcass. Retail stores will take the carcass of the first steer at an advanced price because the retailer can make more from it himself, and because he can dispose of it to a better class of trade. The next paragraph shows how these differences are reflected to the producer.

The Relation of Carcass Price to Hoof Price

The following actual cases taken from animals killed in May, 1920, by Armour and Company show how the demand for different classes of meat is transformed into the value of steers on foot. Only animals of superior breeding and proper finish can make the class of beef represented by carcass No. 1, and the majority of them weigh from 1250 pounds up, although there is no reason why animals of this quality cannot be produced in the cornbelt at 950 pounds for example, when fed from birth. The second class of steer comes more usually from range stock. In the particular instances here quoted, the steers purchased by our buyers to meet the 21 cent trade demand weighed 1400 pounds, while the others weighed 1110 pounds. The data for steers of each class at this time is shown in the following table:

	Steer No. 1	Steer No. 2
Carcass, price	$0.21	$0.185
Carcass weight	840 lbs.	610 lbs.
Value carcass	$176.40	$112.85
Credits—Hides, offal, etc.	40.58	31.41
Killing and overhead	10.50	8.32
Net credits	206.48	135.94
Live weight	1400 lbs.	1110 lbs.
Possible hoof price per cwt	$14.75	$12.25

The actual prices paid were $14.80 and $12.20. From long experience the Dressed Beef Department can figure what it can afford to pay for steers or heifers of any type and weight in order to produce a particular grade or class of beef, the costs being known in terms of averages, and the corrected costs being made for each lot in terms of the actual record of the animals purchased. The figures shown in the preceding table represent corrected costs and not averages used for preliminary estimates. Each morning the Armour cattle buyers are furnished with a statement of the costs of the beef from the animals they purchased the day before, as well as the actual dressing percentage. The principal factor involved in the judgment of the buyer is the ability to estimate the dressing percentage closely, and after years of experience the best buyers become unbelievably accurate. But more than an estimate of the dressing quality is needed, as the buyer must be able to recognize the type of animal that will produce the kind of beef which the Dressed Beef Department needs to fill its orders or its shortages.

PROGRESSIVE BEEF CATTLE RAISING

Part IX

Market Classes of Cattle

How Cattle Are Classified Cattle are placed in classes according to the use to which they are put, while they are graded according to their merit in fulfilling this purpose. The three major classes are *beef cattle*, *butcher stock*, and *feeders and stockers*. Beef cattle produce carcasses suitable for the wholesale trade, of the better grades, Nos. 1 or 2, and of standard quality. Butcher stock produces either an inferior grade of carcass, or else only partially produces marketable cuts. Feeders and stockers are animals that must be developed further before being slaughtered, feeders being ready to go into the feed lot at once, and stockers being too thin or too small to fatten until they have been further developed on cheap feeds. Each of these classes has a certain number of sub-classes, although there is no rigid distinction between them, the daily condition of supply and demand materially affecting the classification. For example light-necked, thin stags on one day's market may be slaughtered as a common grade of stag, while another day they may be sent back to the country as feeders, depending on the relative need for butcher stock or feeders. The principal classes and sub-classes are indicated in the following outline:

Beef Cattle
- Beef Steers
- Yearling Steers
- Yearling Heifers
- Heavy Heifers
- Stags

- Butcher Stock
  - Cows
    - Kosher
    - Butcher
    - Cutter
    - Canner
  - Bulls
    - Butcher
    - Bologna
  - Veals
    - Selected
    - Medium
    - Heavy
- Feeders and Stockers
  - Feeder Steers
  - Yearling Steers
  - Yearling Heifers
  - Feeder Cows
  - Feeder Bulls
  - Springer Cows
  - Springer Heifers
  - Stocker Steers
  - Stocker Heifers

How Cattle Are Graded

Obviously not all of the animals that are placed in a particular class on a given day are equally suited to meet the requirements of that class. Hence they are graded according to their ability to realize these requirements. The standard grades are prime, choice, good, medium, fair, plain, common and poor. *Prime* animals are fully finished and of improved type. *Choice* animals are practically as good in type, but are not so perfectly finished. *Good* animals are not as desirable as prime, either in condition or type. *Medium* steers are practically of the same quality as good, but not their equal in condition, while *fair* steers fall below medium in quality, type and condition. Both medium and fair steers are quite numerous on the market, nearly 50 percent of steers falling in these two grades. *Plain* steers are deficient in type and quality, but carry some flesh, while *common* steers lack flesh to a greater extent. *Poor* steers are typical of their name, inferior in practically all respects.

PROGRESSIVE BEEF CATTLE RAISING

Characteristics of Different Grades and Classes of Beef Cattle and Butcher Stock The following general description of each of the grades of cattle will be found applicable on the average to cattle on the spring Chicago markets of 1920:

Grade	Weight	Condition	Type
Prime Beef Steers	1500–1600	Ripe	Excellent
Choice Beef Steers	1250–1450	Good	Excellent
Good Beef Steers	1250–1450	Good	Good
Medium Beef Steers	1200–1400	Average	Average
Fair Beef Steers	1050–1250	Average	Average
Plain Beef Steers	1000–1150	Fair	Deficient quality and form
Common Beef Steers	900–1050	Light	Few signs of good breeding
Poor Beef Steers	800–950	None	Very inferior

While the other classes and sub-classes vary in weights from those quoted above, the general statements as to type and condition hold good throughout. In some cases a few of the grades are omitted or new grades are created to meet special conditions in a particular class of stock. The following table shows the grades applied to the other classes:

Class	Grades
Yearling Steers	Prime, Choice, Good, Medium, Fair, Plain, Common.
Yearling Heifers	Extra Fancy, Fancy, Prime, Choice, Good, Medium, Fair, Plain, Common.
Heavy Heifers	Extra Fancy, Fancy, Prime, Choice, Good, Medium, Fair, Plain, Common.
Stags	Choice, Good, Medium, Plain, Common, Light, Thin.
Cows	Kosher—Prime, Choice, Good. Butcher—Choice, Good, Fair, Plain. Cutter—Good, Fair, Plain. Canner—Good, Fair, Plain, Common, Poor.
Bulls	Butcher—Prime, Choice, Good, Medium, Plain. Bologna—Choice, Good, Medium, Plain, Common, Light.
Veals	Selected—Prime, Choice. Medium—Choice, Good, Fair, Poor. Heavy—Choice, Good, Fair, Poor.

Extra fancy heifers are females showing very light development in the essentially feminine characters; that is, they are trimmer of middle and smoother through the hooks and rump than usual, being prime in other particulars. They usually are very well bred and extremely uniform. Fancy heifers are similar to extra fancy except that they are slightly less uniform, and usually a little lighter.

Kosher cows are of good size, and well enough finished to make a thick forequarter for the Jewish trade. The quarter is cut off behind the fifth rib, and thickness of meat is essential in order to produce the requirements.

Light thin stags are animals that border between feeders and canners. Their name is indicative of their type and quality.

Prime selected veal calves weigh from 135 to 165 pounds and are fat. Medium weight veals run from 110 to 150 pounds, the better grades being heavier, and the poor to fair veals average from 110 to 125 pounds. Heavy veals weight 200 to 350 pounds, the plain skim milk calves in this class ranging from 200 to 300 pounds.

Grades and Classes of Feeders and Stockers

Feeders and stockers are graded in a manner similar to finished cattle, but the determining factors are based on their ability to gain rather than their ability to kill. Feeder steers have the following classification:

Grade	Weight	Breeding	Type
Fancy selected	1000–1150	Nearly pure beef blood	Uniform, beefy
Choice	1000	High in beef blood	Beefy
Good	900	One or two crosses pure bulls	Above average
Medium	850–900	Mixed	Below average
Fair	800–850	Some cold blood evident	Rougher, plainer

PROGRESSIVE BEEF CATTLE RAISING

Other classes of feeders and stockers grade as shown in the next table, their differences between grades approximating that shown in the foregoing.

Class	Grades	Weights
Yearling steers	Choice, good, fair, common	500–650
Stocker steers	Fancy selected, choice, good, fair, common	600–800
Feeding heifers	Yearlings, choice, good, fair, range	600–800
Feeding cows	Choice, good, fair, plain	650–850
Springer cows	Good, fair	750–900
Springer heifers	Good, fair	700–800
Feeder bulls	Choice, good, fair	800–1100

In the spring of 1920 feeder steers brought about 25 to 50 cents per cwt. more in the corresponding grades than yearlings, due to their ability to finish faster, while stockers sold about a dollar lower than feeders. Heifers in corresponding grades brought from 75 cents to a dollar more than cows, while feeder bulls were generally listed about 25 cents above feeder cows.

In the early days of the cattle industry, feeder and stocker values were set by subtracting the cost of feeding from finished cattle, but as the demand for dressed beef raised the prices of unfinished animals, the margin on which the feeder operates no longer has any relation to the cost of finishing, but is determined by the value of the unfed animal for killing purposes. The feeder buyer frequently finds competition on the highest type of feeder cattle because a limited sale demand exists for just such cuts as the raw feeder produces. The fact that a certain percentage of this type of animals can be used for beef, particularly in the face of market scarcity, has led to a competition with feeder buyers, that has been difficult for them to understand. Many have interpreted this competition to mean that finished cattle are no longer desired, but this is by no means true, since the market can handle only a limited portion of unfinished cattle of this character.

PROGRESSIVE BEEF CATTLE RAISING

Part X

Cattle Types

How Type Is Determined
While finished cattle are classed and graded on the market as indicated in the foregoing paragraphs, there are distinct differences in type between choice *feeders* and choice *killing steers*. The factors that determine whether an animal shall be classed as a beef steer and graded choice include high dressing percent, high proportion of valuable cuts, ability to produce a No. 1 carcass and a size suitable to produce retail cuts most readily marketable. The price paid for live animals is based on these points entirely. On the other hand when a feeder buys a steer he is looking for the points that will indicate profitable utilization of his feed. A steer of this sort has a large rugged frame, a strong chest and constitution, enough depth to indicate a strong feeding capacity, and a loose, mellow, sappy hide that provides a vigorous circulation and a high degree of health. It will be noted that none of the points making the animal profitable as a feeder have any relation to the efficiency with which the steer cuts out, hence the type suitable both to the trade and the feeder is a compromise. This is the type which has come to recognition in the big fat stock shows of England and America, and in its ultimate development provides the show yard champions. The usual champion steer is fed to a flesh unprofitable to the feeder from a market standpoint, since the final gains of such an animal are very costly and there is ordinarily too much fat to permit the animal to be cut up profitably by the butcher. It therefore happens that many times steers gaining high honors in a show are considered less valuable by practical feeders and sell for less on the beef markets than animals of lower show rank.

Characteristics of the Standard Types of Beef Steer

Nevertheless in the long run this compromise type of steer (see page 41) is the one that gives best returns at all stages of his development. Such an animal as viewed from the side should be straight in top and underline, deep, low-set, stylish in carriage, symmetrical in all parts, and possessed of a smooth, thick, meaty appearance. From the rear he should be wide throughout and even; smooth through the shoulders, hook points and rump; and deep and thick in thigh, lower round and twist. From the front he should show a pronounced breadth from shoulder top down through the breast, his neck and shoulder vein should be plump with fat, his head short, broad and well-dished, and his legs set well apart. Such a steer will carry thick cuts in the valuable parts and be proportionate between his carcass and the internal organs that provide his meat making machinery. He should be thick, smooth and mellow to the touch in all parts of his body, and as refined in bone, skin and hair as possible without reducing his ruggedness or vigor. Since beef cattle sell by the pound a big steer at a given age is always preferable to a smaller one of the same general merit.

Dressing Percent

The fat cattle buyer must not only determine what kind of carcass the animal he buys will produce, but he must also determine what the steer will yield, in terms of carcass to live weight. This is known as the dressing percentage and depends on the condition, the freedom from paunchiness, the type and the quality. Fat steers always outdress animals of less finish, the degree of their condition being judged in accurate detail by the filling of the tongue root, brisket, shoulder vein, flank and twist, in addition to the general covering over the body. The fill of the digestive organs with feed and water is as important as the condition. In shipping, steers of 1200 pound weight frequently

PROGRESSIVE BEEF CATTLE RAISING

shrink 40 to 60 pounds, due to the emptying of the digestive tract, which is 3 to 5 percent of the entire weight of the animal. A difference in estimate of 1 percent dress on a 1200 pound steer selling at 15 cents a pound is $1.80, and many mistakes of that sort reduce to zero the usefulness of a buyer. The broad thick type of steer will outdress the steer of wedge-shaped dairy type even when condition and fill are the same, by 3 to 5 percent, while quality in hide, head and bone may affect the dressing ratio by 1 to 2 percent.

The average run of steers killed by Armour and Company dress about 53 percent, good to choice ranging from 56 to 59, and steers of extra good show type, going from 59 to 63. The champion steer at the 1920 Fort Worth show was killed by Armour and Company and dressed 67.48 percent. The world's record is on a spayed heifer killed at the Smithfield Fat Stock Show in London, that made 76.75 percent. Fat cows dress about 56 per cent, and canners from 35 to 43 percent.

PROGRESSIVE BEEF CATTLE RAISING

Part XI

Marketing Cattle

Preparations for Shipping   Much of the profit that may have been acquired during the feeding operation may be lost when the animals are sent to market. Faulty shipping methods may cause such a great difference in the loading weight of the steer at home and its selling weight at the market that the feeder may actually make or break on this margin. This "shrink" is caused by the failure of the animal to eat and drink the normal amount, and by scouring. Long hauls, rough handling, improper feeding, extreme weather, exhaustion and numerous minor factors affect the amount of shrink and nothing will eliminate it entirely. It averages about 4.0 percent of the animal's weight. Grass, silage and pulp-fed cattle shrink more than grain fed, while such grades as canner cows shrink more proportionately than finished stock. If about two days before shipping a less washy ration is substituted for the regular ration, the shrink may not be so great, although if the change is too sudden the animal may be upset. On the other hand, too dry a ration works a severe detriment to the selling condition of the cattle. To withhold water before shipping and to feed salt is cruel and deceives no one except the shipper himself. A normal fill is always accepted but both packer buyers and feeder buyers can easily recognize animals in abnormal condition.

Shipping Counsel   The following suggestions issued by the National Livestock Exchange should prove helpful:

Always route your shipment through to destination and designate each road handling it.

PROGRESSIVE BEEF CATTLE RAISING

Always carefully insert the number and kind of each species of stock loaded.

Always see that car order information is inserted when the car furnished differs from the car ordered.

Always insert the words "ordinary live stock" in the description of stock except that "chiefly valuable for breeding, racing, show purposes or other special uses."

Always insert in the proper space the rate which you understand is to be applied. If the rate and route conflict it is the agent's duty to so inform you.

Always give specific instructions as to the place of feeding enroute, indicating the kind and quantity of feed to be furnished.

Always release your shipment to the 36-hour limit unless, in your opinion, the 28-hour limit should be observed.

Always declare the full value of other than "ordinary live stock," otherwise you cannot recover more than the declared value in case of loss.

Never accept a contract where the carrier's agent seeks to limit the liability of the carrier.

Never declare the value of "ordinary live stock." The agent cannot lawfully require this of you.

Never pay a rate on "ordinary live stock" dependent upon the declared value. If it has been paid file a claim to recover the overcharge.

Never let the railroad agent route your shipment against your own preference. The law gives this right to you exclusively.

Never send an attendant unless he is an experienced livestock man. The responsibilities are too great to risk amateurs.

PROGRESSIVE BEEF CATTLE RAISING

Never pay loading or unloading charges at public markets nor at intermediate feeding stations, except when you order the stock fed there. The law imposes upon the carrier the duty of performing this service.

Always order your car right and in ample time.

Always protect your rights in cases where cars are substituted.

Always bed your car properly.

Always mark your livestock legibly for identification.

Always partition different kinds of livestock and tie dangerous animals.

Always check your railroad billing weight against sale weights to avoid overpayment.

Always pay no more nor no less than the full lawful charge.

Handling Cattle at the Market The *center of consumption* of beef in the United States averages approximately *1100 miles distant* from the *center of production*. As a result of this, a complex but very efficient marketing system has been developed. Cattle shipped to central markets are handled by the trunk railroad, the terminal railroad at the market, the stock yards company and the commission firm before they are manufactured into meat and other products, while the meat passes through the wholesale markets, either directly into the hands of the retailer, or through the intermediate hands of the jobber. Yet, so efficiently is this done in most cases that the retailer can purchase meat more cheaply that has gone through all of these hands and has traveled all of these miles, than he can butcher the animal and sell the meat therefrom himself. Furthermore, by so doing he is insured against disease and can have a greater variety of meats to suit the public taste. The *terminal railroad* receives the cars

PROGRESSIVE BEEF CATTLE RAISING

from the main line, spots them at the chutes for loading and unloading, returns them to the transfer tracks and receives its return by payment of so much per car. The *stock yards company* receives the cattle from the railroad at the unloading chute, counts them, and delivers them to the commission firm at their pens. This company does all the weighing, counts the shipments in the cars and records the entire transaction from unloading to selling. For this, it receives yardage fee. It also furnishes the feed to the shipper for which he must pay. The *commission firm* acts as the selling agent to the shipper. It rents blocks of pens from the yard company and engages in selling or buying the cattle of the shipper. In order to do this, the commission men must know accurately, both cattle and the market, and follow its changes from day to day. Practically no shippers are frequent enough visitors to the market to be able to place as accurate a value on their property as the commission salesman. The commission firm is the final link in the establishment of a cash market since it provides credit for each individual shipper, who is probably unknown to the stockyards company and prepares for him the bill of sale, deducts charges and prepares the check for the shipper often before he has received his own money from the buyer.

Slaughtering Cattle

Cattle bought by Armour and Company for slaughter are driven across to their holding pens. From here they are driven up a long chute or incline to the killing beds on the top floor of the plant. Here they pass into a long line of knocking pens, two to each pen, and are there dispatched with a heavy blow of the sledge. They are then hoisted by the hind legs for sticking, the blood being caught in buckets for use in further manufacture of byproducts, feed and fertilizer. The heads are skinned out, washed and prepared for the Government inspector. The carcass of each individual animal holds its place in rotation throughout the entire

PROGRESSIVE BEEF CATTLE RAISING

operation, all parts cut from it being kept in the same order until the Government inspector has finally passed it. Armour and Company is very proud of the rigid and efficient inspecting force from the Bureau of Animal Industry which supervises the killing and meat preparation in its plants. After the heads have been removed the cattle are moved out from the sticking rail and are laid down on the floor with the feet in the air. The fore and hind legs are skinned out and unjointed at the knee and hock. The legs are sent down another chute to be made into combs, knife handles and glue. The hide is then opened down the center of the belly and skinned off the sides by a set of very expert workmen who with one stroke turn back the hide from the belly to the floor. The cattle are then hooked through the hocks and partially raised from the floor, the middles opened, the entrails removed, placed in a sterilized moving pan and inspected. While this is going on other men skin out the rump and pull the hide free from the round. Extreme care is needed in working here as the hide from the rump makes the very best grade of leather, and any cuts cause serious loss. The tails are skinned out and started on the road to the soup factory. The carcass is then raised completely, the hide removed from the back and "hide droppers" follow to remove the skin entirely from the legs and shoulders. The carcass is now split through the center of the back bone from tail to neck by means of long cleavers and so accurate is the work that jagged cuts and bone splinters are very rare. The split carcass passes on a moving trolley, is given a thorough scrubbing with warm water and brushes, wiped dry and sent white and clean to the last Government inspector. If no disease has been found in any part of the animal it is stamped, "U. S. Inspected and Passed," and sent to the coolers. If infection is found the carcass is switched onto the Government rail and a thorough examination made. Condemned meat is so stamped and kept under Government lock until put in the tank for inedible grease and other by-products. U. S. inspected meat is always safe meat. The Chicago packing plant of Armour and Company can handle 180 cattle at a time, the entire operation from knocking to final inspection taking about an hour. The beef hangs in a temperature of about 34 degrees Fahrenheit for at least 48 hours and is then quartered or cut up otherwise, loaded into the refrigerator car and sent to the branch house for sale to the retailer. Fresh beef is perishable and its handling demands a continuous attention to temperature and a maximum of speed in distribution.

Byproducts When cattle killing first became a centralized business there were only two standard products from the animal marketed, the beef and the hide. Due to the utilization of the byproducts in the modern packing plant, both the dressed meat and the hide bring less than the animal cost on foot, enabling the packer to shave to the lowest degree, the margin between buying and selling price on the cattle he buys and their products. A representative condition *on falling markets* is shown in the following steer, purchased by Armour and Company in May, 1920:

Purchase price, 1,400 pound steer	$207.20	
Selling price (wholesale) 840 pound carcass		$176.40
Selling price, hide		28.10
Killing and overhead costs	10.50	
Credits for raw byproducts		12.48
Loss on steer		.72
Total	$217.70	$217.70

If it were not for the byproducts the loss on this steer would have been $13.20. Armour and Company do not always make a profit per head on their cattle. In 1917 the profit was $1.35 per steer while in 1918 and 1919 an actual loss per animal was sustained.

The possibility of converting these materials which formerly were waste products is based entirely on volume.

PROGRESSIVE BEEF CATTLE RAISING

No small packing plant can afford to organize factories for the manufacture of these materials, but must put in the waste pile everything except what can be most easily assembled. The sources of the byproducts are the hide, the blood, the waste meat, the viscera, the glands and the bones. From the hair and hide come all kinds of leather, brushes, binder for plaster, felt, padding, hair for upholstering and mattresses and glue. From the sinews, fats and blood come bloodmeal, filler for leather, ammoniate for fertilizer, meat meal, lubricating oils, oleomargarine, soap, glue, case hardening bone, gelatine, isinglass and stearine. From the glands and the viscera come goldbeaters' skins, perfume bottle caps, tennis strings, clock cords, drum snares, violin strings, surgical ligatures and pharmaceuticals (such as extract of thyroid, pituitary liquid, pineal substance, suprarenals, pancreatin, adrenalin, pepsin, rennet, thrombo-plastin, etc.) From the bones come combs, buttons, hairpins, umbrella handles, napkin rings, tobacco boxes, buckles, crochet needles, knife handles, dice, chess men, electrical bushings, washers, artificial teeth, bone rings for nursing bottles, glue, case hardening bone, gelatine, fertilizers, oils, grease, soap and red bone marrow. From the hoofs and horns come various manufactured articles of horn, such as inkwells, combs, hair-brush backs, etc., and neatsfoot oil. In the larger manufacturing plants not a single element of what was formerly called "packers' waste" is discarded as of no value.

References

BOOKS ON BEEF CATTLE

"Western Live Stock Management,"
 E. L. Potter (MacMillan & Company).
"Types and Classes of Live Stock,"
 H. W. Vaughan (R. S. Adams & Company).
"Live Stock Judging and Selection,"
 R. S. Curtis (Lea & Febiger).
"Judging Live Stock,"
 John A. Craig (Kenyon Printing & Mfg. Co.,)
"Principles and Practice of Judging Live Stock,"
 Carl W. Gay (MacMillan & Company).
"Cattle Breeds and Management,"
 (Vinton & Company, London).
"Types and Breeds of Farm Animals,"
 C. S. Plumb (Ginn & Company).
"The Breeds of Live Stock,"
 C. W. Gay (MacMillan & Company).
"Shorthorn Cattle,"
 A. H. Sanders (Breeders' Gazette).
"The Story of the Herefords,"
 A. H. Sanders (Breeders' Gazette).
"History of Shorthorn Cattle,"
 MacDonald & Sinclair (Vinton & Company, London).
"Fifty Years With the Shorthorns,"
 Robert Bruce (Vinton & Company, London).
"History of Hereford Cattle,"
 MacDonald & Sinclair (Vinton & Company, London).
"History of Aberdeen-Angus Cattle,"
 MacDonald & Sinclair (Vinton & Company, London).
"Aberdeen-Angus Cattle,"
 A. L. Pulling (Vinton & Company, London).
"Cattle, Breeds and Origin,"
 David Roberts (Dr. David Roberts, Waukesha, Wis.)

BOOKS ON FEEDS

"Feeds and Feeding,"
 Henry & Morrison (The Henry-Morrison Co., Madison, Wis.)
"Productive Feeding of Farm Animals,"
 F. W. Woll (Lippincott).
"The Feeding of Animals,"
 W. H. Jordan (MacMillan & Company).
"First Principles of Feeding Farm Animals,"
 C. W. Burkett (Orange Judd Company).

PROGRESSIVE BEEF CATTLE RAISING

"Profitable Stock Feeding,"
  H. R. Smith (Howard R. Smith, Union Stock Yards, Chicago).
"The Scientific Feeding of Animals,"
  O. Kellner (The MacMillan Company, London).

BOOKS ON BREEDING

"The Principles of Stock Breeding,"
  Jas. Wilson (Vinton & Company, London).
"The Breeding of Animals,"
  F. B. Mumford (MacMillan & Company).
"Breeding Farm Animals,"
  F. R. Marshall (Breeders' Gazette).
"The Breeding of Farm Animals,"
  M. W. Harper (Orange Judd Company).
"Inbreeding and Outbreeding,"
  East & Jones (Lippincott).
"Heredity and Eugenics,"
  W. E. Castle (Harvard University Press).

BOOKS ON DISEASES

"Diseases of Cattle,"
  United States Department of Agriculture.
"Common Diseases of Farm Animals,"
  R. A. Craig (Lippincott).
"Principles of Veterinary Science,"
  F. B. Hadley (Saunders).

Publications of the United States Department of Agriculture, available for free distribution by the Department:

"Lespedeza or Japan Clover," Farmer's Bulletin No. 441.
"Red Clover," Farmer's Bulletin No. 455.
"Market Hay," Farmer's Bulletin No. 508.
"Vetches," Farmers' Bulletin No. 515.
"Crimson Clover," Farmers' Bulletin Nos. 550, 579 and 646.
"Making and Feeding of Silage," Farmers' Bulletin No. 578.
"Beef Production in the South," Farmer's Bulletin No. 580.
"Economical Cattle Feeding," Farmer's Bulletin No. 588.
"Sudan Grass," Farmers' Bulletin No. 605.
"Breeds of Beef Cattle," Farmers' Bulletin No. 612.
"Cottonseed Meal for Feeding," Farmers' Bulletin No. 655.
"Field Peas," Farmers' Bulletin No. 690.
"Stock Losses from Poisonous Plants," Farmers' Bulletin No. 720.
"Natal Grass," Farmers' Bulletin No. 726.
"Contagious Abortion," Farmers' Bulletin No. 790.
"Production of Baby Beef," Farmers' Bulletin No. 811.
"Home-made Silos," Farmers' Bulletin No. 855.
"Dehorning and Castration," Farmers' Bulletin No. 949.
"Growing Beef on the Farm," Farmers' Bulletin No. 1073.

Unloading chutes for cattle at the stock yards.

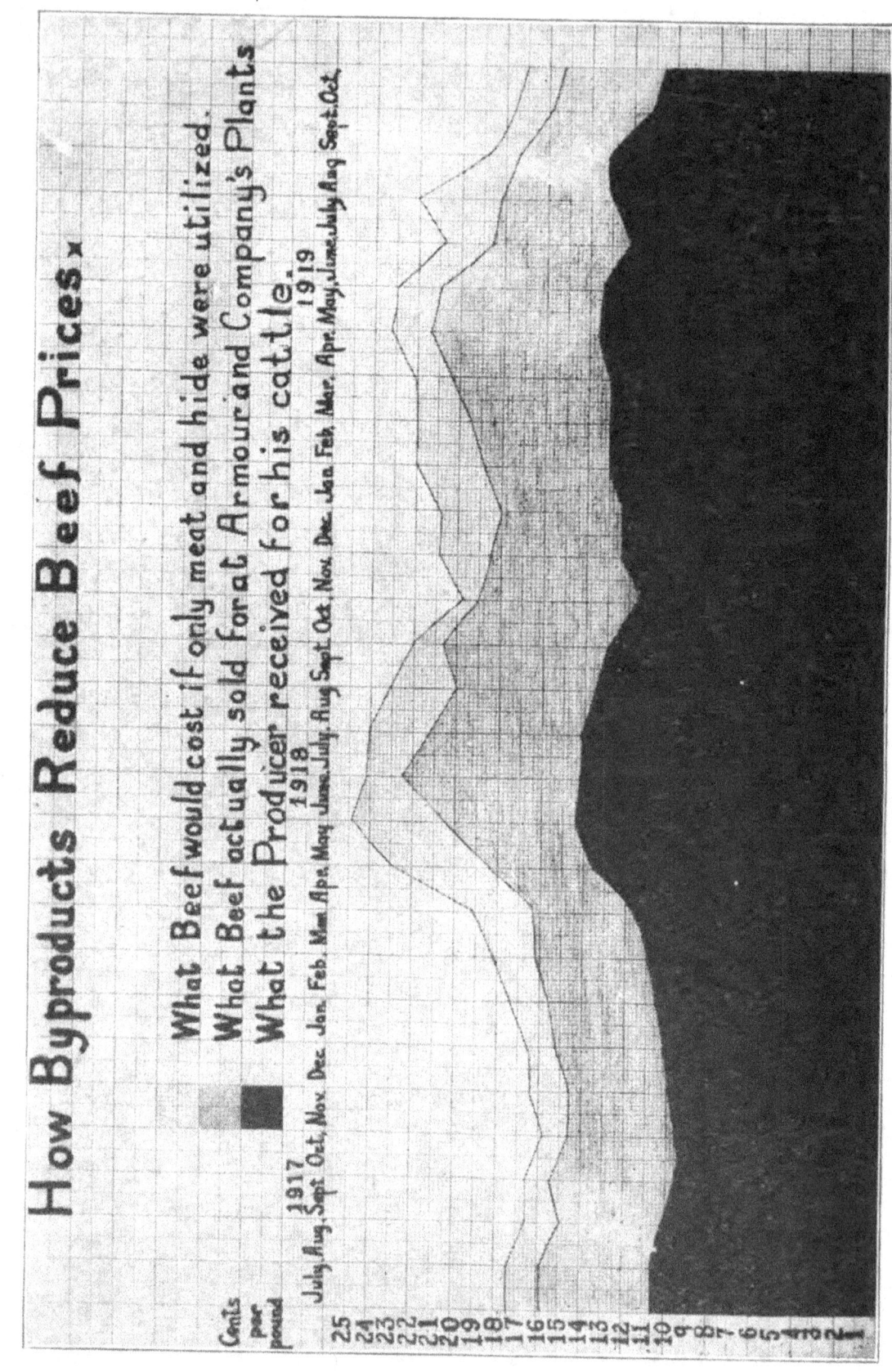

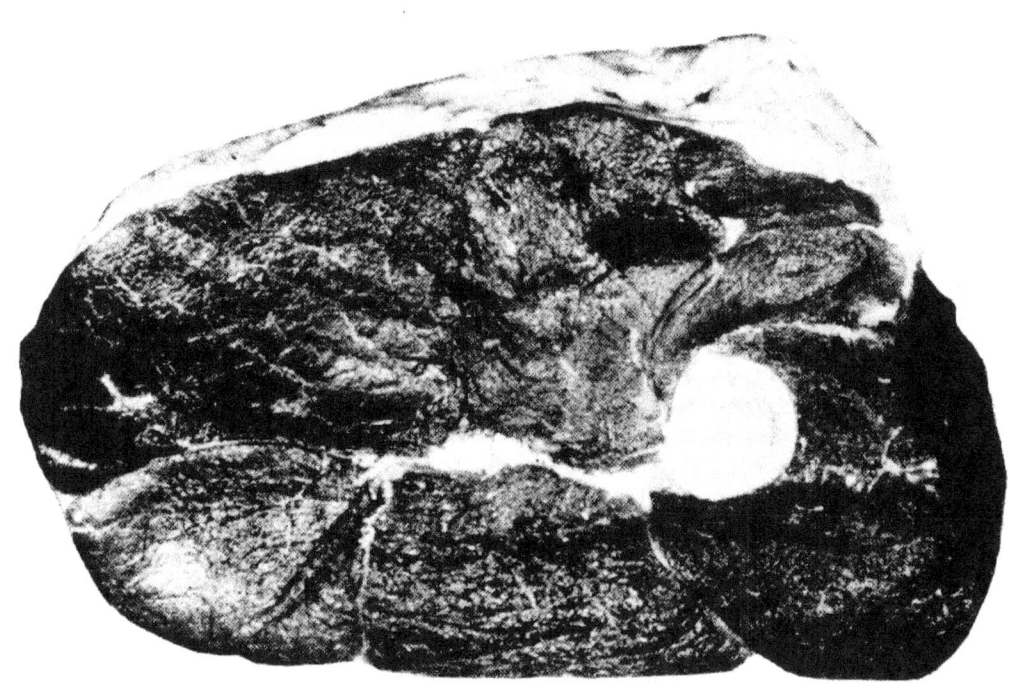

No. 3 round.

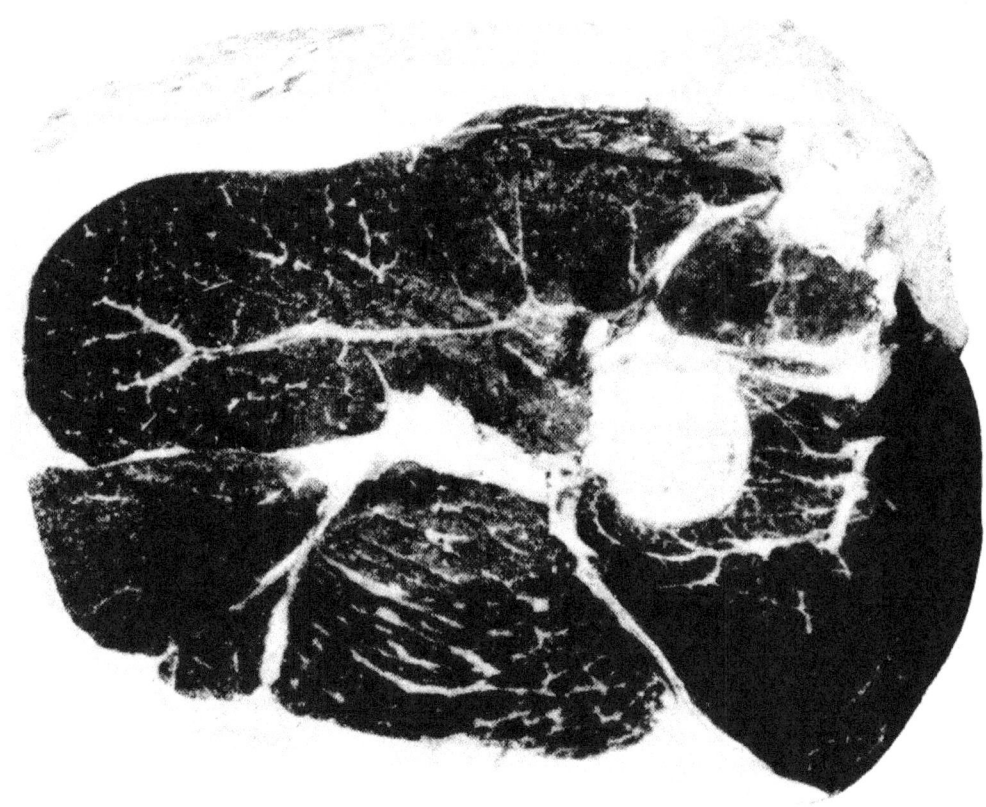

No. 1 round.

Note the fuller shape of the No. 1 round and the better marbling of fat with lean. Also the surface of the No. 1 round is velvety and dry as compared with the darker wetter surface of the No. 3 round.

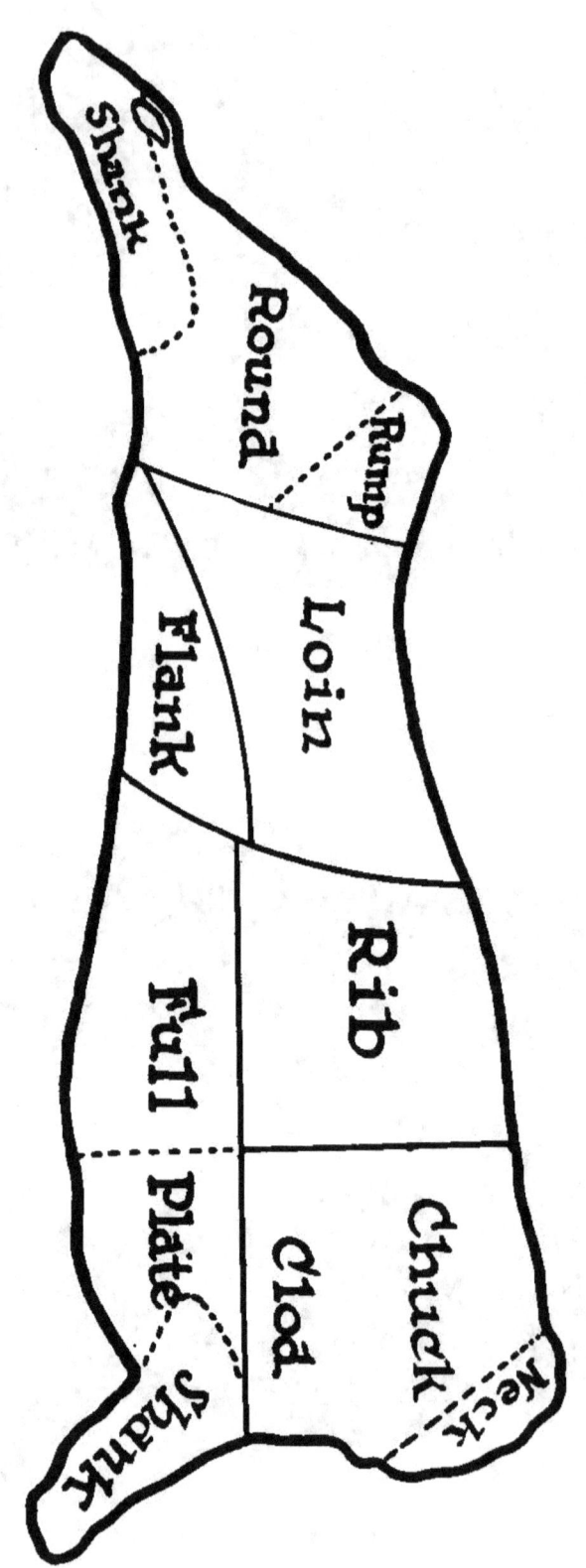

Wholesale cuts of the beef carcass.

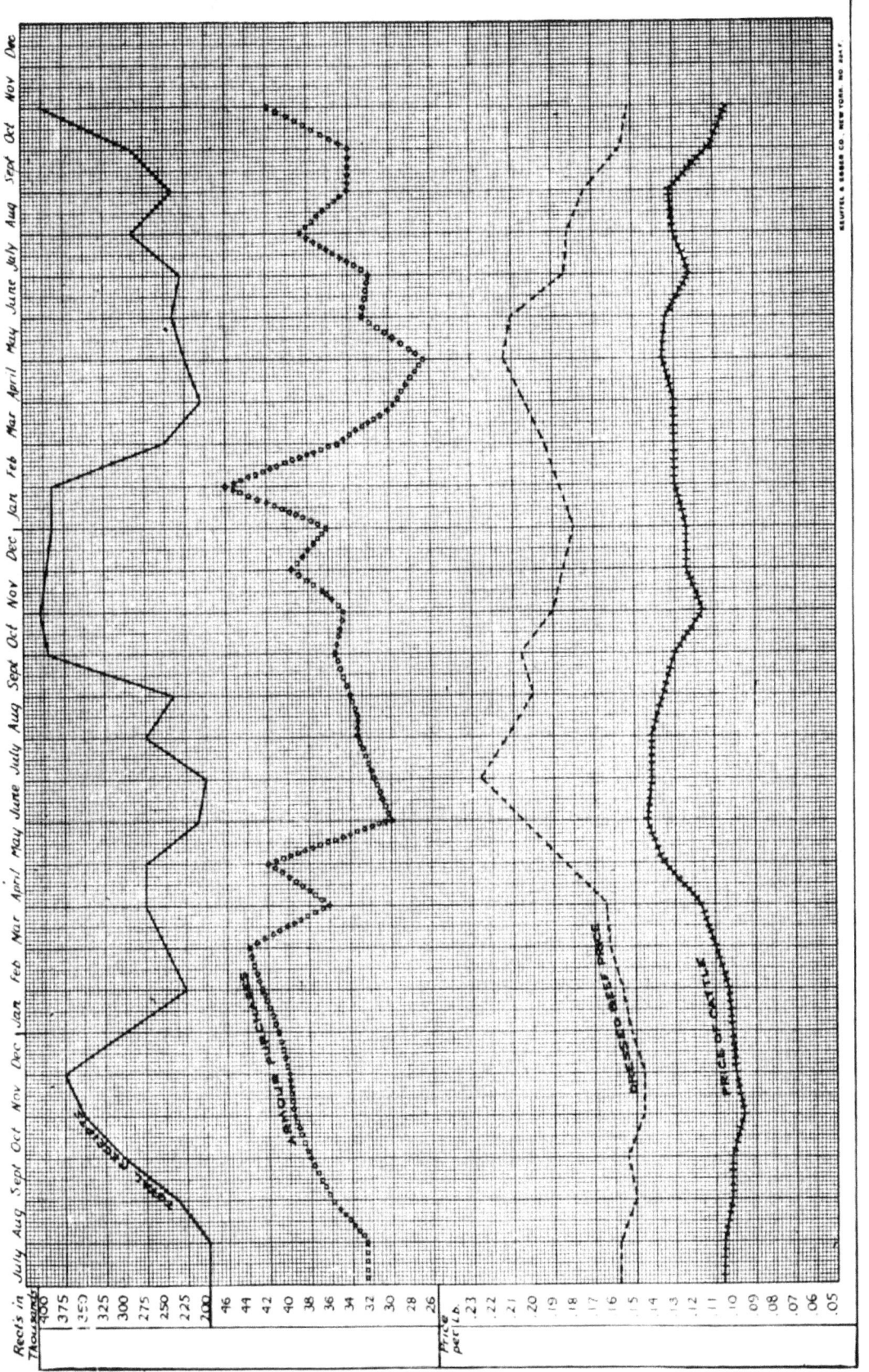

Curve showing relative effect of supply and demand as illustrated by receipts and wholesale price of beef on the price of live cattle, from July 1917 to November 1919.

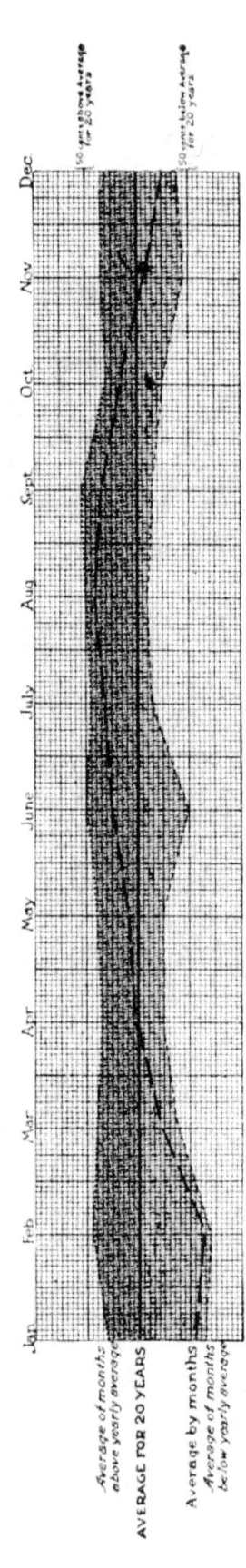

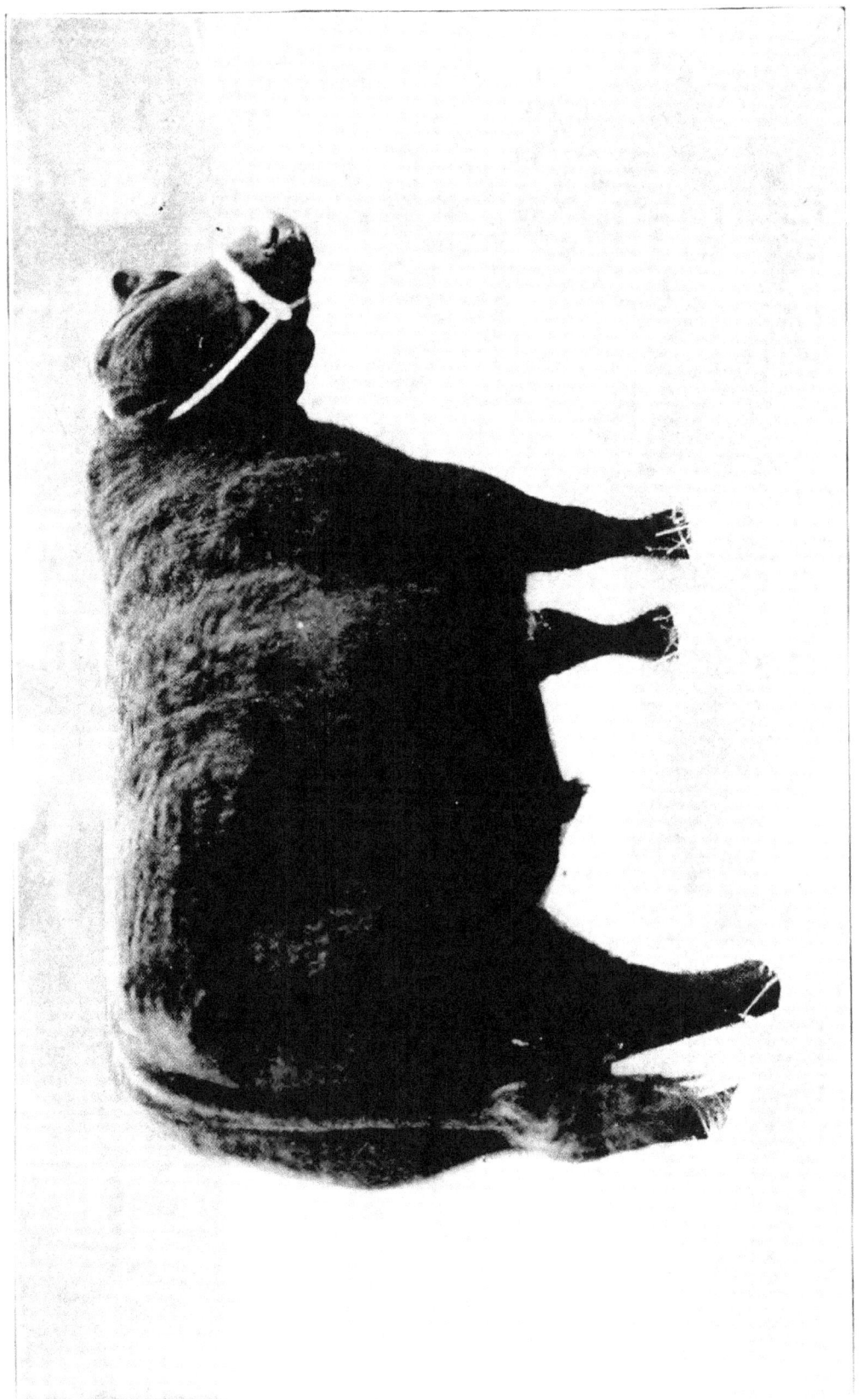

The standard type of beef steer, "Victor," 1911 International grand champion. (See Page 62.)

Tubercular cow apparently healthy in appearance which reacted to the tuberculin test and was found diseased on killing. (See page 39.)

A steer showing a pronounced case of lump jaw. (See page 37.)

The water hole.

A tile silo representative of one of the many types of permanent construction adopted throughout the country.

A strictly beef type, the Shorthorn heifer "Lady Supreme." (See page 10.)

A draft, beef and milk type of France, Simmenthal-Fribourgeois cross.
(See page 10.)

A beef and milk type, the champion Red Polled cow "Constant."

Steers are thriftiest when fed in the open.

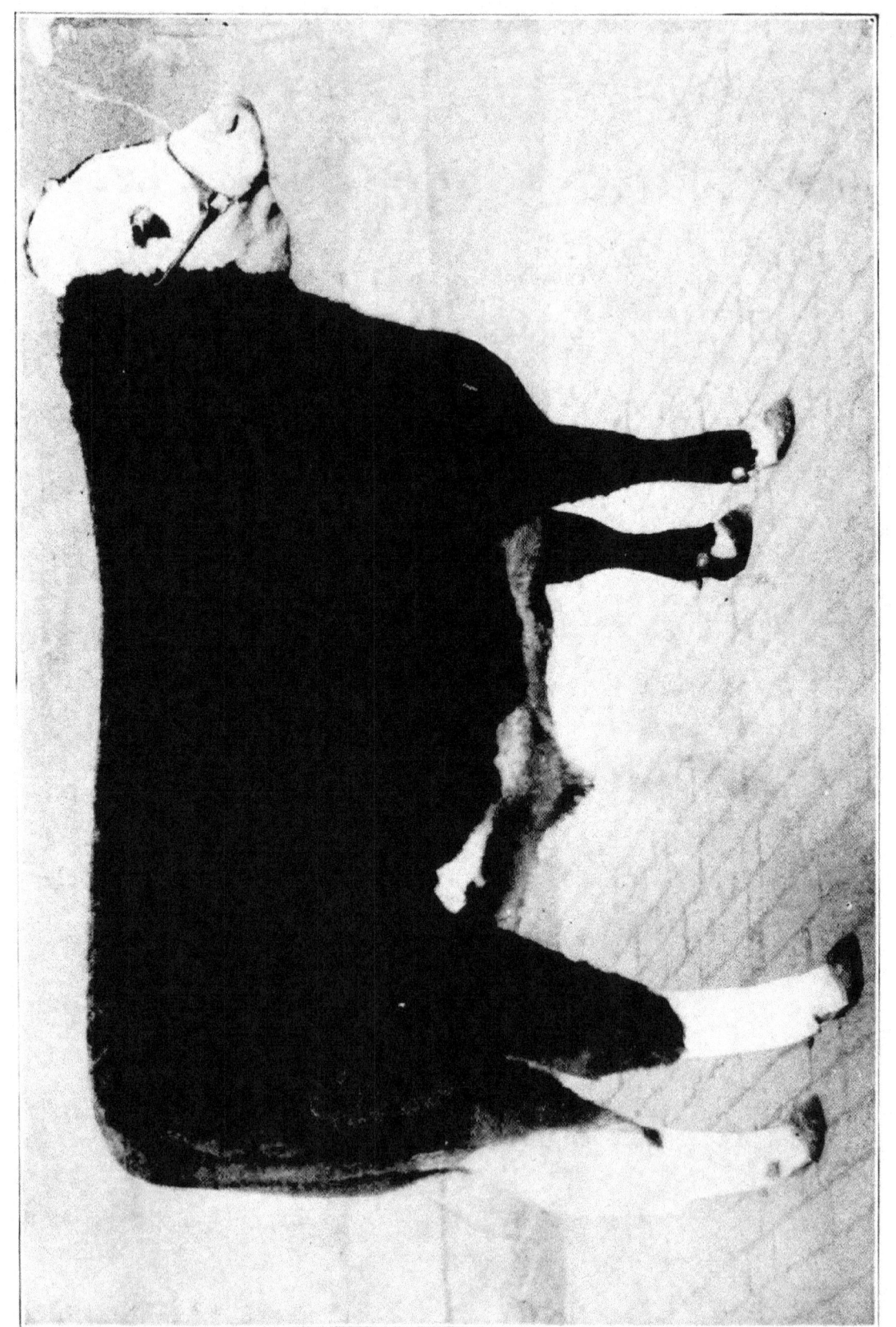

The cross-bred Aberdeen-Angus Hereford steer "California Favorite," illustrating transmission of traits from both sire and dam. The white markings of the Hereford have been transmitted as well as the polled trait and black color of the Aberdeen-Angus. The conformation is intermediate between the Hereford and Aberdeen-Angus.